T0210779

EVOLUTIONARY ROBOTICS:
FROM ALGORITHMS TO IMPLEMENTATIONS

WORLD SCIENTIFIC SERIES IN ROBOTICS AND INTELLIGENT SYSTEMS

Editor-in-Charge: C J Harris (*University of Southampton*)

Published:

World Scientific Series in Robotics and Intelligent Systems – Vol. 28

EVOLUTIONARY ROBOTICS: FROM ALGORITHMS TO IMPLEMENTATIONS

Lingfeng Wang

Texas A&M University, USA

Kay Chen Tan
Chee Meng Chew

National University of Singapore, Singapore

World Scientific

NEW JERSEY · LONDON · SINGAPORE · BEIJING · SHANGHAI · HONG KONG · TAIPEI · CHENNAI

Published by

World Scientific Publishing Co. Pte. Ltd.

5 Toh Tuck Link, Singapore 596224

USA office: 27 Warren Street, Suite 401-402, Hackensack, NJ 07601

UK office: 57 Shelton Street, Covent Garden, London WC2H 9HE

Library of Congress Cataloging-in-Publication Data
Wang, Lingfeng.
    Evolutionary robotics : from algorithms to implementations / Lingfeng Wang, Kay Chen Tan, Chee Meng Chew.
        p. cm. -- (World Scientific series in robotics and intelligent systems ; v. 28)
    Includes bibliographical references and index.
    ISBN-13 978-981-256-870-0
    ISBN-10 981-256-870-0
    1. Evolutionary robotics. I. Tan, Kay Chen. II. Chew, Chee Meng. III. Title. IV. Series
TJ211.37 .W36 2006
629.8'92--dc22

                                        2006285523

British Library Cataloguing-in-Publication Data
A catalogue record for this book is available from the British Library.

First published 2006 (Hardcover)
Reprinted 2016 (in paperback edition)
ISBN 978-981-3203-32-7

Copyright © 2006 by World Scientific Publishing Co. Pte. Ltd.

All rights reserved. This book, or parts thereof, may not be reproduced in any form or by any means, electronic or mechanical, including photocopying, recording or any information storage and retrieval system now known or to be invented, without written permission from the Publisher.

For photocopying of material in this volume, please pay a copying fee through the Copyright Clearance Center, Inc., 222 Rosewood Drive, Danvers, MA 01923, USA. In this case permission to photocopy is not required from the publisher.

Printed in Singapore

To our families, for their love and patience.

Preface

Modern robotics has moved from the industrial manufacturing environment to human environment for service and entertainment in the recent few years. The human environment is typically unstructured and quite often ever-changing, the robot is required to be capable of independently learning the dynamic environment as well as continuously adjusting its behavior to accomplish the desired tasks during its execution. It should be able to capture the properties of the environment effectively and make suitable decisions accordingly to properly deal with the current situation in real-time. In order to accomplish the desired tasks, mobile robots should have a certain level of intelligence to deal with the uncertainties occurred the environment that they are operating in. For instance, if a mobile robot is required to keep track of a moving object, it needs to determine by itself the movement of the target and plan its appropriate path to closely follow the target. Meanwhile, the robot may be asked to execute other tasks such as obstacle avoidance during its motion. However, the pre-programmed robots can not accomplish such tasks adaptively since they have no ability to handle the real operating situations in a smart fashion.

In the design of intelligent artifacts, robot engineers have always been inspired by nature. For instance, the design of these intelligent systems has being influenced significantly by the physiology of natural evolution. The simplicity and robustness of biological species are the highly desired characteristics for autonomous robotic systems. Like their biological counterparts, autonomous mobile robots should be able to perceive both static and dynamic aspects of the external environment and adjust their behaviors accordingly in order to adapt to it. Fortunately, nature provides invaluable inspirations for the design of robotic system capable of exhibiting certain intelligent behaviors. Inspired by the biological solution, we attempt to build a variety of intelligent behaviors for the autonomous mobile robot. In the research reported in this book, a thorough literature review, followed by representative and extensive experiment studies, illustrates the effectiveness of the biologically inspired approach to intelligent robotic controller design. The emphasis of this book is placed on the artificial intelligence based evolutionary robotics.

There are two major methods in controller design for intelligent robots, one of which is hardware-based design and another is software-based (i.e., algorithm-based) design. For the hardware-based robotic controller design, robotic controller is derived at the evolvable hardware level, which is a novel and salient set of integrated circuits capable of reconfiguring their architectures using artificial evolution techniques. In this book, two chapters are dedicated to the evolvable hardware based robotics. Software (or algorithm)-based design is more conventional with respect to the hardware-based one. In this designs, the emphasis is put on the design and implementation of software-based robotic controllers. The controllers are usually compiled and burned in the processor which is embedded in the autonomous robotic systems. In this book, five real-world examples of software-based robotic controller designs are discussed in detail.

Structure of the Book. The primary motivation for this book is to discuss the application of biologically-inspired approaches in the design and development of autonomous intelligent robotic systems. A variety of techniques were employed to implement the different robotic behaviors. Except for the evolvable hardware based evolutionary robotics, other design techniques used for autonomous intelligent robotic systems are also discussed in the book, which include fuzzy logic, evolutionary computing, reinforcement learning, and so forth. The book presents a very detailed review of the state-of-the-art in the domain of evolutionary robotics and some case studies on practical design and implementation of intelligent robotic systems are also fleshed out. The overall book structure is arranged in the following way: In the first chapter, the basic concepts regarding artificial evolution and evolutionary robotics are introduced, and then a variety of successful applications of artificial evolution in autonomous robot navigation along the dimension of artificial evolution adopted are surveyed and discussed. Open issues and future research in this field are also presented. The second chapter surveys the application of evolvable hardware in evolutionary robotics, which is an emerging research field concerning the development of evolvable robot controller at the hardware level in order to adapt to dynamic changes in the environment. The context of evolvable hardware and evolutionary robotics is reviewed respectively, and a few representative experiments in the field of robotic hardware evolution are presented. As an alternative to conventional robotic controller designs, the potentialities and limitations of the evolvable hardware based robotic system are discussed and summarized. The third chapter presents the design and real-time implementation of an evolvable hardware based autonomous robot navigation system using intrinsic evolution. Distinguished from the traditional evolutionary approaches based on software simulation, an evolvable robot controller at the hardware gate-level that is capable of adapting dynamic changes in the environments is implemented. The fourth chapter presents the design and implementation of an autonomous robot navigation system for intelligent target collection in dynamic environments. A feature-based multi-stage fuzzy logic sensor fusion system is developed for target recognition, which is capable of mapping noisy sensor inputs into reliable decisions. The robot exploration and path planning are based on a grid map oriented reinforcement path learning system, which allows for long-term predictions and path adaptation via dynamic interactions with physical environments. The fifth chapter presents a new approach of task-oriented developmental learning for humanoid robotics. It is capable of setting up multiple tasks representation automatically using real-time experiences, which enables a robot to handle various tasks concurrently without the need of predefining the tasks. The sixth chapter discusses a general control architecture for bipedal walking which is based on a divide-and-conquer approach. Based on the architecture, the sagittal-plane motion-control algorithm is formulated using a control approach known as virtual model control. A reinforcement learning algorithm is used to learn the key parameter of the swing leg control task so that stable walking can be achieved. In the seventh chapter, a genetic algorithm tuned fuzzy logic controller is proposed for bipedal walking control implementation. The basic structure of fuzzy logic controller is constructed based on the linear inverted pendulum model. Genetic algorithm is implemented to search and optimize the fuzzy logic controller parameters. The eighth chapter investigates the application of genetic algorithm as an optimization tool to search and optimize key parameter in the walking controlling of a humanoid robot. Virtual model control is employed as a control framework where ankle gain plays an important part in regulating forward velocity during walking. The final chapter concludes the book and discusses the possible future research directions in this field.

Readers. The communities of intelligent control and autonomous robotics in both academia and industry should find that this book is useful and able to provide an up-to-date view of the fields. It is hoped that this book can serve as a good reference for further study as well as provide a different perspective on robot control. The utility of the entire AI-based approach is however open to debate and ultimately will only show its strength if it stands the rest of time, or, at least in the short term, provides sufficiently effective and convincing solutions to current challenging real-world problems. The book is likely to be useful to people in the areas of intelligent systems, autonomous robots, intelligent robotics, or what has been called "evolutionary robotics" as indicated in the book title. While this book will likely be of most interest to electrical engineers, mechanical engineers, and computer scientists, it may also be interesting to persons in other fields such as biological sciences due to its interdisciplinary feature.

The reader is assumed to be familiar with the fundamental knowledge in algorithm, software, and hardware designs. In industry, practitioners, evolutionary designers, and hardware designers will probably like the research results reported in this book. In academia, teachers and students can utilize any individual chapters as the reference materials for their lectures and studies in evolutionary algorithms, evolutionary robotics, and evolvable hardware.

Acknowledgments. Many people deserve special acknowledgment for their help throughout the entire duration of research, book writing, and production. First, the assistance from Dr. Christopher J. Harris, the Editor-in-Charge of World Scientific Series in Robotics and Intelligent Systems, is highly appreciated. We also wish to thank the people who communicated with us in World Scientific Publishing Inc. and Imperial College Press, especially to Mr. Steven Patt (Desk Editor) and Mrs. Katie Lydon (Editorial Assistant), for all their kind help in the acquisition, editing, and production processes of this book. For the preparation of the first three chapters, we would like to express great thanks to Dr. K. Ide in the Applied AI Systems, Inc., Canada and Dr. A. Thompson at the University of Sussex, U. K. for their invaluable comments and encouragement at the stage of evolvable hardware study. Our special thanks go to O. Carmona, S. Legon, and A. Griffith in the K-team, Inc., Switzerland for their timely technical support. Dr. T. Hirst at the Open University in U. K. and Dr. H. Sakanashi at the ETL of Japan also helped us a lot by providing useful literature. All of the work reported in the first five chapters was conducted in the Center for Intelligent Control (CIC), Department of Electrical and Computer Engineering, National University of Singapore. The work of R. Xian, Y. J. Chen, P. Xiao, and X. Liu laid solid foundations for the research reported in these chapters. For Chapter 6, the authors wish to thank Dr. Gill A. Pratt for his comments provided to this research. For Chapter 7, the authors would like to thank Min Du for helping to carry out the implementation of the simulation and Weiwei Huang for helping to organize the data. For Chapter 8, the authors wish to thank Professor G.-S. Hong for his valuable comments given to this project. We also like to thank the graduate students, H.-N. Ho and T. Sateesh, for working on the simulation implementation, designing the robot's hardware, and writing the draft report for this research. We are also indebted to many colleagues and friends for their assistance and advice. Our friends have stood by and encouraged us when our productivity waned. Finally, we wish to thank our families for their ever-present encouragement and for the moral and practical support given over the years before and throughout this endeavor. We dedicate this book to them.

The readers are encouraged to send us their questions arisen in reading the book. Any problems and comments can be directed to the first author at l.f.wang@ieee.org. We hope readers enjoy the volume! Go forth and produce!

L. F. Wang, K. C. Tan, and C. M. Chew

Contents

Chapter 1

Artificial Evolution Based Autonomous Robot Navigation

Modern robots are required to carry out work in unstructured dynamic human environments. In the recent decades, the application of artificial evolution to autonomous mobile robots to enable them to adapt their behaviors to changes of the environments has attracted much attention. As a result, an infant research field called evolutionary robotics has been rapidly developed that is primarily concerned with the use of artificial evolution techniques for the automatic design of adaptive robots. As an innovative and effective solution to autonomous robot controller design, it can derive adaptive robotic controllers capable of elegantly dealing with continuous changes in unstructured environments in real time. In the chapter, the basic concepts regarding artificial evolution and evolutionary robotics are introduced, and then a variety of successful applications of artificial evolution in autonomous robot navigation along the dimension of artificial evolution adopted are surveyed and discussed. Open issues and future research in this field are also presented.

1.1 Introduction

Early robots were nothing more than clever mechanical devices that performed simple pick-and-place operations. Nowadays robots are becoming more sophisticated and diversified so as to meet the ever-changing user requirements. The robots are developed to perform more precise industrial operations, such as welding, spray painting, and simple parts assembly. However, such operations do not really require the robot to have intelligence and behave like human beings since the robots are simply programmed to perform a series of repetitive tasks. If anything interferes with the pre-specified task, the robot cannot work properly anymore, since it is not

1

capable of sensing its external environment and figuring out what to do independently.

Robotics today has moved from the structured factory floors to the unpredictable human environment [Khatib, Brock, and Change, et al., 2002]. Therefore, traditional manipulator controlled robots are being replaced by the emerging autonomous intelligent mobile robots. Let's first look at what capabilities a robot should have:

Autonomous: having autonomy; not subject to control from outside; independent (Webster's New Universal Unabridged Dictionary, Barnes and Noble Books, New York, 1996).

Intelligent: pertaining to the ability to do data processing locally, smart (Webster's New Universal Unabridged Dictionary, Barnes and Noble Books, New York, 1996).

Such robots have the ability to adjust their behaviors autonomously in the ever-changing physical environment. Here a simple definition of a robot could be "a mechanism which is able to move and react to its environment". There are many types of robots and the robots discussed in this chapter are autonomous mobile robots. Autonomous robots refer to the agents capable of executing the specified tasks without human intervention by adjusting their behaviors based on the real environment. Mobile robots refer to those which navigate and perform tasks without external intervention. Thus, an autonomous mobile robot should be able to make decisions independently and adaptively in the real-world environments. As an example, complex robotic tasks such as trash collection using autonomous robots can be broadly applied to a variety of fields such as product transferring in manufacturing factory, rubbish cleaning in office, and bomb searching on battle field, etc. Such robots should be able to cope with the large amount of uncertainties existing in the physical environment. For instance, the robot may be required to achieve certain goals without colliding with obstacles in this motion. However, the physical environment is usually dynamic and unpredictable. Quite often the obstacles are not static and moreover, the sensor readings are imprecise and unreliable because of the noises. Therefore, the robot cannot operate properly in the real world anymore with a highly pre-programmed controller. For robots to become more efficient, maintenance free, and productive, they must be capable of making decisions independently according to the real situations [Staugaard, Jr., 1987]. The robot must be able to adjust its behavior by itself online and make appropriate decisions under various uncertainties encountered during its motion in an independent and smart fashion.

Therefore, a key challenge in autonomous robotics is to design control algorithms that allow robots to function adaptively in unstructured, dynamic, partially observable, and uncertain environments [Sukhatme and Mataric, 2002]. Humanoid robots [Brooks, 2001, 2002; Lund, Bjerre, and Nielsen, et al., 1999], self-reconfiguring robots [Rus, Kotay, and Vona, 2002], probabilistic robotics [Thrun, 2002], entertainment robotics [Veloso, 2002] are some representatives in this emerging field these years.

In the last several decades, biologically inspired robotics [Taubes, 2000; Full, 2001] such as evolutionary robotics (ER) [Floreano, 1997; Nolfi, 1998b] is being developed as an innovative research field, which is an artificial evolution based methodology to obtaining self-adaptation and self-learning capabilities for robots to perform desired missions. For instance, investigations are being conducted on the use of robots in the imitation of life [Holland, 2001] and the integrated design using a co-evolutionary learning approach [Pollack, et al., 2001].

Several researchers have given overviews on evolutionary robotics and specific topics such as neural networks, fuzzy logic, and evolutionary algorithms several years ago [Nolfi, Floreano, and Miglino, et al., 1994; Gomi and Griffith, 1996; Walker and Oliver, 1997; Meyer, Husbands, and Harvey, 1998; Meeden and Kumar, 1998; Wang, 2002]. This review is intended to give a more state-of-the-art discussion, emphasizing the artificial evolution based approach to autonomous robot navigation in terms of methods and techniques currently used, and open issues and future trends, most of which are not covered in previous ER surveys. The remainder of the chapter is organized as follows. Section 1.2 presents the characteristics of evolutionary robotics. The adaptive autonomous robot navigation is discussed in Section 1.3, where online and offline evolutions are compared. In Section 1.4, some typical artificial evolution methods used in deriving robotic controllers are discussed one by one, which include neural networks, evolutionary algorithms, fuzzy logic, and other methods. Open issues and future prospects are detailed in Section 1.5, which include SAGA, combination of evolution and learning, inherent fault tolerance, hardware evolution, online evolution, and ubiquitous and collective robots. Finally, the conclusion is drawn in Section 1.6.

1.2   Evolutionary Robotics

Robots have developed along two paths, i.e., industrial and domestic. Industrial robots have been developed to perform a variety of manufacturing tasks such as simple product assembly. However, most industrial robots are unintelligent as they cannot hear, see, or feel. For instance, if assembly parts are not presented to the robot in a precise and repetitive fashion, the robot cannot perform its task properly. In autonomous robot navigation, the robot should be able to move purposefully and without human intervention in environments that have not been specifically engineered for it. Autonomous robot navigation requires the robot to execute elementary robotic actions, to react promptly to unexpected events, and to adapt to unforeseen changes in the environment.

To become more intelligent in robot navigation, an autonomous robot should be able to sense its surroundings and respond to a changing environment promptly and properly. For instance, in the robotic task of target collection, target recognition and path planning are the two challenging tasks that a robot should execute. Target recognition under uncertainties is difficult to implement because the sensor information is usually too noisy to be directly mapped to a definitive target representation. Multi-sensor data fusion has turned out to be an effective approach to resolve this problem. Sensor data fusion is to combine and integrate inputs from different sources into one representational format. It yields more meaningful information as compared to the data obtained from any individual sources. Data fusion in robotics is very often used in applications such as pattern recognition and localization for autonomous robot navigation. The associated techniques include the least square method (LSM), Bayesian method, fuzzy logic, neural network, and so on. Each of these methods has their own merits. For instance, fuzzy logic has the advantage of mapping imprecise or noisy information as input into the output domain. It is particularly effective in autonomous robot navigation since the data is obtained under uncertain conditions. Their operation and inference rules can be represented by natural languages in form of linguistic variables. Traditional robot controllers are often designed to run in a well-defined environment to execute certain repetitive or fixed actions. However, actual autonomous robotic applications may require the mobile robots to be able to react quickly to unpredictable situations in a dynamic environment without any human intervention. For instance, personal and service robots are two important trends for the next generation of robotic applications [Schraft and Schmierer, 2000]. In both cases, what matters most is the necessity to adapt to new and unpredictable situations, react quickly, display behavioral robustness, and operate in close interaction with human beings [Floreano and Urzelai, 2000; Hopgood, 2003]. However, for such mobile robot applications, it is difficult to model unstructured external environments perceived via sensors in sufficient detail, which are changing continuously. Therefore, traditional robotic controllers designed for factory automation are not suited for these types of applications anymore. The designed robot should be able to have the capability of self-organization for navigating in such dynamic environments. Self-organizing systems can operate in unforeseen situations and adapt to changing conditions. There are a variety of robotic architectures when building autonomous robot navigation systems. The most commonly used one is the hierarchical architecture, where functionalities of the robotic system are decomposed into high- and low-level layers [Nicolescu and Mataric, 2002]. The high-level layer is responsible for system modeling and planning and the low-level layer is in charge of sensing and execution. The second type of robotic architecture is called behavior-based architecture [Sridhar and Connell, 1992; Laue and Rfer, 2004; Aravinthan and Nanayakkara, 2004]. The complicated robotic behavior is obtained by assembling some more solvable and simpler robotic behavior components. Another robotic architecture is the hybrid architecture [Connell, 1992; Secchi, et al., 1999]. It is the hybrid of layered organization and behavior-based decomposition of the execution layer. Behavior-based robotic system design is a commonly employed approach in building autonomous robot navigation systems. We need to enable the robot to autonomously coordinate behaviors in order to exhibit the complicated robotic behavior, e.g., coordinating target-tracking and anti-collision behaviors in order to reach a target position while avoiding unforeseen and moving obstacles during its motion. It offers a collection of fundamental robotic behaviors and the behavior is selected according to its response to a certain stimulus in the real situations. In behavior-based robotics, basic robotic behaviors are elementary building blocks for robot control, reasoning, and learning. The environment plays a central role in activating a certain basic robotic behavior throughout the execution of robotic systems. In general, both the robotic behavior modules and the coordination mechanisms are designed through a trial-and-error procedure. By doing so, the designer gradually adjusts them and examines corresponding behaviors until the satisfactory robotic behavior is derived. Inspired by the Darwinian thinking, evolutionary algorithms [Rechenberg, 1973; Holland, 1975] are being applied to the robotic control field [Gomi and Griffith, 1996; Meyer, Husbands, and Harvey, 1998]. In this discipline, control algorithms inspired largely by biological evolution are used to automatically design sensorimotor control systems. Evolutionary robotics (ER) is a novel and active domain, where researchers apply the evolution mechanism to the robot design and control. The fundamental principles of evolution mechanism include a population of individuals competing for survival, inheritance of advantageous genes from the parent generation and the offspring variability possibly result in a higher chance of survival. ER enables robotic system to adapt to unknown or dynamic environments without human intervention. The robotic controllers derived by evolutionary methods have advantages of relatively fast adaptation time and carefree operations [Floreano and Mondada, 1996; Lund and Hallam, 1997; Nolfi and Floreano, 2000]. In evolutionary robotics, an initial population of chromosomes is randomly generated, where each chromosome encodes the controller or morphology of a robot. Each physical or simulated robot then acts guided by the robotic controller, and in the evolutionary process, the robotic performance in various situations is evaluated. The fittest robots have more chances to generate copies of their genotypes using genetic operators such as reproduction, crossover, and mutations. This process is iterated until an individual with satisfactory performance is attained. That is, artificial evolution starts with a population of strings or genotypes, which represents the control parameters of a robot. Each robot is evaluated in the problem domain and given a fitness value. Following the inspiration from Darwinian evolution, the fittest genotypes have a better chance to breed, which usually includes crossing two parents to produce a child and altering or mutating random values. As the evolution process proceeds, mean fitness within the task domain increases. Thus, evolutionary robotics is completely different from the conventional robotic controller design, where the desired robotic behavior is decomposed into some simpler basic behaviors. Under the framework of the evolutionary algorithm-based robotic controller, the sensory inputs are encoded into the genetic strings and sent to the evolutionary algorithm based robotic controller. The robotic controller makes appropriate decisions based on the real-time sensory values and sends its decision to the execution devices such as motors and actuators. Most commonly, the sensory signals are encoded into a binary string, where each bit represents corresponding sensory information. For instance, in the binary string, the first bit may indicate the status of the frontal sensor value of the mobile robot. "1" indicates it detects a front obstacle and "0" tells that there is no obstacle in front of the robot. This encoding mechanism can also be applied to the encoding of behavioral block. Each binary bit indicate whether or not its corresponding basic behavioral block is used in responding to the current real-world situations. The evolutionary algorithm performs a parallel search in the space of Boolean functions, searching for a set of functions that exhibits the required robotic behaviors in the real-world environment. It generates new Boolean functions by transforming and combining the existing functions using various operators in a genetically inspired way. Statistics related to their performance are used to steer the search by choosing appropriate functions for transformation and combination. For instance, a standard genetic algorithm with Roulette-wheel selection, single-point crossover, and random mutation can be used in the robotic controller evolution. When all the individuals of a population have been examined, the genetic operators are used to produce a new generation of individuals. Roulette-wheel selection includes a linear scaling of the fitness values and a probabilistic assignment for each offspring individual according to its fitness. All offspring are then randomly paired and a random single-point crossover is performed based on the pre-specified crossover probability. Then the newly obtained strings are mutated segment by segment with the assigned mutation probability. From the robotic point of view, the mutation operator is able to introduce some new robot motions for sensory input states and find appropriate motions. The mutation rate must be carefully chosen and a tradeoff between evolution stability and convergence should be reasonably made. For instance, if the mutation rate is too high, it will incur genetic noise and may even cause evolution instability. If it is too low, the robot may take a much longer time to acquire the expected behaviors. In general, the mutation rate used is slightly higher than the inverse of the length of the chromosomes. In the overall genetic evolution process, the elitist strategy is quite useful in the selective reproduction operation, which ensures that the best chromosome is reserved in the next generation. The elite strategy prevents the best chromosomes in each generation in case that the experimenter is not sure whether or not the specified mutation rate consistently brings noise into the evolution process.

Other main techniques for autonomous robotics include artificial life [Brooks, 1992], reinforcement learning [Tan, et al., 2002b], artificial immune system [Fukuda, et al., 1999], and plan-based control [Beetz, 2002; Beetz, et al., 2002], and so on. In the remainder of this chapter, we primarily discuss the evolutionary robotics and its related issues.

1.3   Adaptive Autonomous Robot Navigation

The perception-action cycles are too complicated to be systematically formulated using a unified theory. A perception-action cycle usually involves a large amount of highly interacting components (i.e., interrelated subsystems), that continuously interact with each other during the perception-action cycle. The interactions for generating the aggregate behavior are highly nonlinear, so that the aggregate behavior cannot be derived easily from the behaviors of isolated components using traditional mathematical formulism. The components in a perception-action cycle are highly diverse and their behaviors are closely associated with each other. For instance, malfunction of one component may cause a series of changes and adjustments on other components. And it is possible to compensate for the damaging effect caused by the failed component during the reception-action cycle via self-repairing capability. This type of self-reconfiguration in the reception-action process is not easy to be clearly represented by classical mathematical approaches. Furthermore, the internal models of components, the large amount of subtly interwoven information processing components, and their high inherent nonlinearities, make traditional approaches incapable of formulating the complex system dynamics systematically. Perception-action cycle will still remain an open problem unless we are able to take all of the aforementioned factors into consideration simultaneously. As a result, there is still no a unified theory for the perception-action cycle at the moment.

Generally speaking, the approaches to adaptive autonomous robotic navigation can be classified into two categories [Jakobi, 1997]. The first is to evolve the system in a pre-designed environment where various sensor readings needed for training the desired robot behavior are intentionally designed (a.k.a. off-line evolution) normally via a simulator [Seth, 1998]. The fitness evaluation can be performed by simulation where different controllers can be tested under the same environment conditions. A simulation is usually cheaper and faster than a real robot. The individuals might exploit simplifications or artifacts in the model. However, the evolved robotic controller might not work reliably on the real robot due to the simplified environment model used. The second scheme is to conduct evolution during the operational phase so that the robot behavior can be gradually improved by continuous learning from the external environment (a.k.a. on-line evolution) [Yao and Higuchi, 1999]. The fitness is evaluated using the real robot in the real world where fitness evaluation is a stochastic process. In addition to the robot learning system, a changing environment must be taken into consideration. In addition, the robot may encounter unexpected situations which do not appear in simulations. Therefore, it is expected that the evolved robotic controller is more robust than that obtained by simulation.

However, both approaches are not well established as of today and there are still many unsolved problems. For instance, the adaptive behavior acquired from the first method is not strong enough to handle the unpredictable human environment. In essence, such behaviors can only work well in the predictable environment. Once a trained robot encounters an environment that it has-not seen in its training phase, the robot may get stuck and take undesired actions. In real-world applications, it is not possible to take all the feasible events into account during the evolution process. Or even assuming it were possible, the training cost would be extremely high. The second method would require certain new opportunities for improving the robot behavior during the operational stage. In this method, the environment intended for the robot training should be carefully designed so that there are complete feasibilities for the new behaviors to be tried out. However, one of the main drawbacks of this training mode is that it is extremely time-consuming. The time taken to obtain a target controller in on-line evolution is primarily determined by fitness evaluation process for a particular task. It is difficult to evaluate individuals on-line due to the large number of interactions with the physical environments. In particular, since actions to be taken in the current sensor input state affect the future sensor state, it is difficult to present regular sensor input patterns to each variant controller in an efficient manner. If the problem cannot be well resolved, the fitness function approach applied to off-line hardware evolution cannot be used in the on-line evolution situations.

For instance, in the hardware based evolution system [Tan, et al., 2002a], various methodologies were investigated in the early stage of the study to find out the most suitable approach for the hardware evolution system, including the method of on-line intrinsic evolution. The obstacle blocks are densely distributed in the enclosed field so that the robot has higher chances to interact with the obstacles and different sensor input states could be presented to the robot in an efficient manner. For each chromosome, the reflected infrared is measured twice, i.e., one for sensor input states and the other for sensor-based fitness evaluations. Instead of assigning fitness value based on Field Programmable Gate Array (FPGA) outputs, the robot moves in the physical environment for 0.3 second and evaluates the chromosome based on the new sensor states. The desired behaviors lead to a state where the robot faces fewer obstacles than in the previous state. The difference in the sensor states is used to calculate the fitness score for each chromosome. However, the on-line fitness evaluation turned out to be impractical because too much time is required for the evaluation process. Therefore, on-line evolution considerably decreases the efficiency of the evolutionary approach, due to the need of a huge number of interactions with the environment for learning an effective behavior.

Two physical experiments were conducted to find solutions to this dilemma. In the first experiment, a random motion was carried out after each fitness evaluation via the FPGA output values (the wheel speeds). Such a random motion may generate a new world state, which is different from previous world states for the current chromosome. In the evolution, six classified input states need to be shown to the robot requesting for its response, and the fitness evaluation is then carried out. Therefore, many random motions are needed to search for different input states for each chromosome, and a lot of time is consumed for such an exhaustive state search that causes the evolution to be unpredictable.

The second experiment is similar to the first one in principle, but adopts a different strategy to search for the input states. In this approach, instead of looking for all the defined input states for consecutive chromosomes, algorithm is applied to search for the defined input states for all chromosomes at each generation. Once duplicated input state is presented to the robot, the next chromosome that has not encountered the current input state will take over the current chromosome for fitness evaluation. Such an approach seems to evaluate all the chromosomes in each generation at a much faster speed than the previous method. However, it is still not capable of coping with the massive data needed to process in our application, and a lot of time is wasted in searching for the sensor input states rather than in the intrinsic evolution.

As seen from the above attempts to conduct online evolutions, it is hard for the robot to achieve the adaptation in the operational phase, i.e., online evolution, though this method exhibits the real idea about adaptive robot evolutions.

In autonomous robot navigation applications, the reinforcement learning mechanism [Mitchell, 1997; Sutton, 1998; Sutton and Barto, 1998] is very often used for path planning (e.g., to figure out the shortest route to the designated destination via unsupervised learning). In doing so, the robot is capable of planning navigation path based on the acquired sensor values coupled with certain appropriate self-learning algorithms instead of a predefined computational model, which is normally not adaptive. In reinforcement learning, situations are mapped to actions based on the reward mechanism. The autonomous robotic system takes an action after receiving the updated information collected from the environment. It then receives a reward as an immediate payoff. The system purposefully maintains the outputs that are able to maximize the received reward over time. In doing so, the cumulative reward is maximized. The four basic components in reinforcement learning are the policy, reward, value, and model. According to Sutton [Sutton and Barto, 1998], policy is the control strategy which determines how the agent chooses the appropriate actions to achieve its goals. Reward reflects the intrinsic desirability of the state and the reward accumulated over time should be maximized. The reward mechanism can be provided internally or externally. The state value is defined as the total reward it receives from that state onwards, which indicates what types of actions are desired in the long term. Finally, the real-world environment can always be seen as the collection of uncertain events. A predefined model is not capable of adapting to an unpredictable environment. Therefore, based on the real-time sensory values, the robot should be able to maintain and update the world model autonomously and adaptively during its movement. In the learning process of an autonomous mobile robot, it judges if it is desirable to take a certain action in a given state using trial and error search and delayed reward. Reinforcement learning enables an autonomous mobile robot to sense and act in its environment to select the optimal actions based on its self-learning mechanism. Two credit assignment problems should be addressed at the same time in reinforcement learning algorithms, i.e., structural and temporal assignment problems. The autonomous mobile robot should explore various combinations of state-action patterns to resolve these problems.

1.4 Artificial Evolution in Robot Navigation

It has shown in [Nolfi, 1998a] that the robot behaviors could be achieved by the simpler and more robust evolutionary approaches than the traditional decomposition/integration approach. This section gives an overall introduction of the artificial evolution mechanism. It presents the main strategies for robotic controller design. Various applications of artificial evolution in robotics are surveyed and classified. Furthermore, in this section their specific merits and drawbacks in robotic controller design are discussed. At present, there is little consensus among researchers as to the most appropriate artificial evolution approach for heterogeneous evolutionary systems. It should be noted that the section only gives some basic ideas on what artificial evolution is due to the space restriction and the chapter theme. For more detailed information about these topics, readers are suggested to refer to specific references such as [Goldberg, 1989; Fogel, 2000a, 2000b; Driankov and Saffiotti, 2001; Tan, et al., 2001a, 2001b].

1.4.1   *Neural Networks*

Many evolutionary approaches have been applied to the field of evolvable robotic controller design in the recent decades [Meyer, 1998; Chocron and Bidaud, 1999; Pollack et al., 2000]. Some researchers used artificial Neural Networks (NN) as the basic building blocks for the control system due to their smooth search space. NNs can be envisaged as simple nodes connected together by directional interconnects along which signals flow. The nodes perform an input-output mapping that is usually some sort of sigmoid function. An artificial NN is a collection of neurons connected by weighted links used to transmit signals. Input and output neurons exchange information with the external environment by receiving and broadcasting signals. In essence, a neural network can be regarded as a parallel computational control system since signals in it travel independently on weighted channels and neuron states can be updated in parallel. NN advantages include its learning and adaptation through efficient knowledge acquisition, domain free modeling, robustness to noise, and fault tolerance, etc. [Huang, 2002]. The behaviors that evolutionary robotics is concerned with at present are low-level behaviors, tightly coupled with the environment through simple, precise feedback loops. Neural networks are suitable for this kind of applications so that the predominant class of systems for generating adaptive behaviors adopts neural networks [Jakobi, 1998]. The same encoding schemes can be used independently of the specific autonomous robot navigation system since different types of functions can be achieved with the same type of network structure by varying the properties and parameters of simple processing used. Other adaptive processes such as supervised and unsupervised learning can also be incorporated into NN to speed up the evolution process.

NNs have been widely used in the evolutionary robotics due to the aforementioned merits. For instance, locomotion-control module based on recurrent neural networks has been studied by Beer and Gallagher [1992] for an insect-like agent. Parisi, Nolfi, and Cecconi [1992] developed back propagation neural networks for agents collecting food in a simple cellular world. Cliff, Harvey, and Husbands [1993] have integrated the incremental evolution into arbitrary recurrent neural networks for robotic controller design. Floreano and Mondada [1996] presented an evolution system of a discrete-time recurrent neural network to create an emergent homing behavior. Using a dynamic evolution system, Steels [1995] proposed a coupled map latticed to control a robot. Lund and Miglino [1996] evolved neural network control system for the Khepera robot using a simulator. A modified back-propagation algorithm for development of autonomous robots was proposed by Murase et al. [1998] to control the real Khepera robot. Smith [1998] used feed-forward neural network to teach a robot to play football. A two stage incremental approach was used to simulate the evolution of neural controllers for robust obstacle avoidance in a Khepera robot by Chavas et al. [1998]. By analyzing the fitness landscape of neural network which controls a mobile robot, Hoshino, Mitsumoto, and Nagano [1998] have studied the issue on the loss of robustness of evolved controllers. In [Ishiguro, Tokura, and Kondo, et al., 1999], the concept of dynamic rearrangement function of biological neural networks is incorporated with the use of neuromodulators to reduce the robotic controller performance loss between the simulated and real environments. Hornby et al. [2000] built a recurrent neural network based robotic simulator that runs at over 11000 times faster than real time and used it to evolve a robust ball-chasing behavior successfully. Spiking neural controllers were evolved for a vision-based mobile robot [Floreano and Mattiussi, 2001]. Reil and Husbands [2002] used recurrent neural networks to deal with the control problem of bipedal walking.

However, neural networks also have certain drawbacks. For instance, a NN cannot explain its results explicitly and its training is usually time-consuming. Furthermore, the learning algorithm may not be able to guarantee the convergence to an optimal solution [Huang, 2002].

1.4.2 *Evolutionary Algorithms*

There are currently several flavors of evolutionary algorithms (EAs). Genetic Algorithms (GAs) [Holland, 1975] is the most commonly used one where genotypes typically are strings of binary. Genetic Programming (GP) [Koza, 1994] is an offshoot of GAs, where genotypes are normally computer programs. Other flavors such as Evolution Strategies (ES) are also used in

ER. Many concerns are shared among these approaches.

In principle, GA is a simple iterative procedure that consists of a constant-size population of individuals, each one represented by a finite string of symbols, known as the genome, encoding a possible solution in a given search space, which consists of all possible solutions to the target problem. Generally, the genetic algorithm is applied to space which is too large to be exhaustively searched. The symbol alphabet can be binary encodings, character-based encodings, real-valued encodings, and tree representations. The standard GA proceeds as follows: an initial population of individuals is created randomly or heuristically. At every evolutionary step (i.e., a generation), the individuals in the current population are decoded and evaluated based on some pre-specified fitness criterion (i.e., fitness function). To generate a new population, individuals are chosen according to their fitness. Many selection procedures can be used in deriving robotic controller. For instance, the fitness-proportionate selection is the simplest one, where individuals are selected based on their relative fitness. By doing so, the individuals with high fitness stand a better chance to reproduce while the ones with low fitness are prone to die.

The majority of GA work is search problems in a fixed-dimensional search space. The well-defined finite dimensionality of the search space allows a choice of genotype coding with fixed length. Typically, the GA starts with a population of random points effectively spanning the entire search space. Successive rounds of genetic operations lead to the optimum or near optimal population. In autonomous robot navigation, we assumed that the desired robot behaviors such as anti-collision could be accomplished with the evolved optimal chromosome. However, this assumption may not be true since we could not obtain the complete anti-collision behavior using the best chromosome achieved from evolution. The best solution may lie outside the boundaries that we defined in the experiments. Therefore, other new approaches need to be adopted in order to achieve complete obstacle avoidance behavior. There are two alternatives for solving this problem: incremental learning and species adaptation genetic algorithm (SAGA). Incremental evolution is able to derive a sequence of increasingly complex tasks. This is particularly useful when trying to resolve a complex and difficult problem. Genotype lengths are allowed to change in SAGA, which is discussed in the next section.

As a commonly used EA, GA has also been used in [Husbands, Harvey, and Cliff, 1995; Floreano and Mondada, 1996] for generating robotic behaviors. Based on GA techniques, Yamada [1998] trained the robot to recognize uncertain environments. Gaits of a lagged robot were derived by the GA in [Gomi and Ide, 1998]. Barfoot and D'Eleuterio [1999] used GA to implement the multi-agent heap formation. Using perception-based GA, Kubota et al. [1999] realized collision avoidance of the mobile robot in a dynamic environment. Chaiyaratana and Zalzala [1999] presented the use of neural networks and GAs to time-optimal control of a closed-loop robotic system. Earon et al. [2000] developed walking gaits based on cellular automation for a legged robot where GA was used to search for cellular automata whose arbitration results in successful walking gaits. Inspired by the prior work using GA, Thompson [1995] adopts the conventional GA as the training tool to derive the robot controllers in the hardware level. The encouraging experimental results justify the effectiveness of GA as a robust search algorithm even in hardware evolution.

Most applications nowadays use the orthodox GA, however, Species Adaptation GAs (SAGA) proposed by [Harvey, 1992, 2001] would be more suitable for certain robot evolution applications such as evolvable hardware based robotic evolutions. In SAGA, different structures are encoded with genotypes of different lengths, which offers a search space of open-ended dimensionality. Cyclic Genetic Algorithm (CGA) was also introduced in [Parker, 1999] to evolve robotic controllers for cyclic behaviors. Also distributed genetic algorithms are introduced into the evolutionary robotics field these years. For instance, in the spatially distributed GA, for each iteration a robot is randomly selected from a population distributed across a square grid. The robot is bred with one of its fittest neighbors and their offspring replaces one of the least fit neighbors such that the selection pressure keeps successful genes in the population. The distributed GA is usually robust and efficient in evolving capable robots. GA exhibits its advantages in deriving robust robotic behavior in conditions where large numbers of constraints and/or huge amounts of training data are required [Walker and Oliver, 1997]. Furthermore, GA can be applied to a variety of research communities due to its gene representation. However, GA is computationally expensive [Walker and Oliver, 1997]. Though GA is now being widely used in ER field, a variety of issues are still keeping open in the GA-based ER. For instance, the fitness function design is an important issue in GA-based evolution schemes [Revello and McCartney, 2000]. The fitness function should be its measurement of its ability to perform under all of the operating conditions. In principle, all these objectives can be fulfilled by setting an appropriate fitness function so as to derive the desired robotic performance exhibited during autonomous navigation. Therefore, the fitness function design needs to be investigated more carefully to make the robot evolve in a more effective way. Several experiments have also been performed where the robotic controllers were evolved through Genetic Programming (GP) [Koza, 1994; Dittrich, Burgel, and Banzhaf, 1998]. Brooks [1992] has outlined a GP-based evolutionary approach applied to lisp-like programming languages. Reynolds [1994] used GP to create robot controllers that enabled a simple simulated vehicle to avoid obstacles. Nordin and Banzhaf [1997] reported experiments using GP to control a real robot trained in real environments with actual sensors. Later they proposed a new technique allowing learning from past experiences that are stored in memory. The new method shows its advantage when perfect behavior emerges in experiments quickly and reliably [Nordine, Banzhaf, and Brameier, 1998]. Liu, Pok, and Keung [1999] implemented a dual agent system capable of learning eye-body-coordinated maneuvers in playing a sumo contest using GP techniques. An example-based approach to evolve robot controllers using GP was implemented in [Lee and Lai, 1999]. A GP paradigm using Lindenmayer system re-writing grammars was proposed as a mean of specifying robot behaviors in the uncertain environment [Schaefer, 1999]. Kikuchi, Hara, and Kobayashi [2001] implemented a robotic system featuring reconfigurable morphology and intelligence using GP.

1.4.3   *Fuzzy Logic*

As mentioned earlier, fuzzy logic provides a flexible means to model the nonlinear relationship between input information and control output [Hoffmann, Koo, and Shakernia, 1998]. It incorporates heuristic control knowledge in the form of if-then rules, and is a convenient alternative when the system to be controlled cannot be precisely modeled [Paraskevopoulos, 1996; Driankov and Saffiotti, 2001]. They have also shown a good degree of robustness in face of large variability and uncertainty in the parameters. These characteristics make fuzzy control particularly suited to the needs of autonomous robot navigation [Saffiotti, 1997]. Fuzzy logic has remarkable features that are particularly attractive to the hard problems posed by autonomous robot navigation. It allows us to model uncertainty and imprecision, to build robust controllers based on the heuristic and qualitative models, and to combine symbolic reasoning and numeric computation. Thus, fuzzy logic is an effective tool to represent real world environments. In evolutionary robotics, fuzzy logic has been used to design sensor interpretation systems since it is good at describing uncertain and imprecise information.

All the specific methods have their own strengths and drawbacks. Actually they are deeply connected to one another and in many applications some of them were combined together to derive the desired robotic controller in the most effective and efficient manner. For instance, Fuzzy-genetic system [Hagras, Callaghan, and Colley, 2001] is a typical evolution mechanism in evolving adaptive robot controller. Arsene and Zalzala [1999] controlled the autonomous robots by using fuzzy logic controllers tuned by GA. Pratihar, Deb, and Ghosh [1999] used fuzzy-GA to find obstacle-free paths for a mobile robot. Driscoll and Peters II [2000] implemented a robotic evolution platform supporting both GA and NN, Xiao, et al. [2002] designed autonomous robotic controller using DNA coded GA for fuzzy logic optimization.

1.4.4 *Other Methods*

Apart for the above commonly used methodologies, several other evolutionary approaches were also tried out in the ER field in recent years. For example, classifier systems have been used as an evolution mechanism to shape the robotic controllers [Dorigo and Schnepf, 1993; Dorigo and Colombetti, 1994]. Grefenstette and Schultz used the SAMUEL classifier system to evolve anti-collision navigation [Grefenstette, 1989; Grefenstette and Schultz, 1994]. Katagami and Yamada [2000] proposed a learning method based on interactive classifier system for mobile robots which acquires autonomous behaviors from the interaction experiences with a human. Gruau and Quatramaran [1997] evolved robotic controllers for walking in the OCT-1 robot using cellular encoding. In the work of Berlanga et al. [1999], the ES is adopted to learn high-performance reactive behavior for navigation and collisions avoidance. Embodied evolution was proposed as a methodology for the automatic design of robotic controllers [Watson, Ficici, and Pollack, 1999], which avoids the pitfalls of the simulate-and-transfer method.

Most of the ER approaches aforementioned are essentially software-based. More recently, hardware-based robotic controllers using artificial evolution as training tools are also being exploited. The development of evolvable hardware (EHW) has attracted much attention from the ER domain, which is a new set of integrated circuits able to reconfigure their architectures using artificial evolution techniques unlimited times. Higuchi, Iba, and Manderick [1994] used off-line model-free and on-line model-based methods to derive robot controllers on the logic programmable device. Attempting to exploit the intrinsic properties of the hardware, Thompson [1995] used a Dynamic State Machine (DSM) to control a Khepera robot to avoid obstacles in a simple environment. The hardware evolution issues will be detailed in the next section.

1.5   Open Issues and Future Prospects

Although a variety of successes have been achieved in ER, there are still a bunch of issues to be addressed and there also exists plenty of room to further explore the relevant research areas. For instance, Lund, Miglino, and Pagliarini, et al. [1998] intended to make evolutionary robotics an easy game, and Mondada, and Legon [2001] explored interactions between art and mobile robotic system engineering. In this section, open issues as well as future prospects for evolutionary robotics are presented.

1.5.1   *SAGA*

As discussed earlier, it is highly difficult to develop a fitness evaluation strategy in standard GA to progressively improve the robot behavior. The EAs such as conventional GA and GP are not suitable for such incremental evolutions. SAGA [Harvey, 1992, 2001] extends the traditional GA to attack this issue, where mutation is the primary genetic operator and crossover does not play the driving force as it does in conventional GAs. SAGA is more appropriate than orthodox GA for certain ER applications such as hardware-based robot evolution.

As mentioned in the earlier sections, it would be difficult to develop a fitness evaluation technique that allows a progressive path of improvements starting from an initial random population and finally reaching a complex target behavior. The EAs such as conventional GA and GP are not suitable for this long-term open-ended incremental evolution of complex systems. A slightly more complex problem might not be solved properly using the same experimental parameters since it requires several more generations. The final population suited for the original problem could not be used as the starting population of the next complex task to evolve in an incremental manner. Harvey's SAGA [1992] was developed as an extension of the GA to cope with this issue. In SAGA, mutation is the primary genetic operator while crossover is not as crucial as it is in conventional GAs in the evolution process.

SAGA should be more appropriate than standard GA for the autonomous robotic application. There is no pre-specified chromosome length in the structure to be designed in autonomous robot navigation. In SAGA, different structures are encoded with genotypes of different lengths. This makes the evolutionary search operate in the space of varying dimensionality. Progress though such a genotype space is accomplished by adjusting the genotype length. Meanwhile, it should be noted that except for the genotype length, many other factors should also be carefully considered to attain the optimal string configuration for the desired robotic performance.

1.5.2 *Combination of Evolution and Learning*

Evolution and learning are two forms of biological adaptation that differ in space and time. Evolution is a strategy for a population of robots (physical or simulated) to adapt to the environment through genetic operations in a distributed population of individuals. However, learning is an adaptation strategy for a robot to adapt to its environment by conducting a set of modifications in each individual during its own lifetime. Evolution is a form of adaptation capable of capturing relatively slow environmental changes that might encompass several generations. Learning enables an individual to adapt to dynamic and unpredictable environmental changes at the generation level. Learning includes a variety of adaptation mechanisms, which are able to produce adaptive changes in an individual level during its lifetime. Learning and evolution are highly related to each other. Learning allows individuals to adapt to changes in the task environment that occur in the lifespan of an individual or across few generations and enables evolution to use information extracted from the environment thereby channeling evolutionary search. It is a replenishment of evolution and can help and guide evolution. However, it may involve high computational cost. For instance, sometimes robots based on reinforcement learning can be very slow in adapting to environmental changes and thereby may incur delays in the ability to acquire fitness and even system unreliability. For instance, reinforcement learning allows an autonomous robot that has no knowledge of a task or an environment to learn its behavior by gradually improving its performance based on given rewards in performing the learning task.

1.5.3 *Inherent Fault Tolerance*

It is also observed that not all device properties are constant over time as many factors may change as time goes on during system operations, which may incur the internal failure of the robot. A practical and useful robot controller should be able to operate properly in face of temperature changes and supply voltage fluctuations. Also it should be able to work in the presence of medium instability. A fit robotic behavior must be capable of handling the full variety of conditions that it may encounter in the practical environment. In the robotic applications, the entire autonomous robot can be viewed as an evolvable system since it is able to adjust its own structure and strategy. In light of the variations in operating conditions, the robot should have the ability to perform the desired tasks as before.

Meanwhile, the ability to embrace the inevitability of the external robot failure is also crucial to guarantee proper autonomous robot navigation in harsh environments. The evolved robot should also be able to detect any fault that occurs at any time in the sensor-motor system and respond fast enough to cover up the fault. That is also the purpose of implementing the on-line evolution in ER evolutions. Embedding fine-grained Micro-Electro-Mechanical Systems (MEMS) components into robotic systems can bring stronger processing capacities. However, sensor and actuator failures may often lead to performance deterioration or dysfunction of robotic systems. In most cases, it is not known when and how much a component fails. To guarantee the proper operations of the overall robotic system, the remaining components should be reconstructed to compensate for the damaging effects caused by the failed components in a timely manner. As a result, an effective fault tolerance based robotic controller for such component failure compensation in a real-time fashion is highly desirable.

Another issue in fault tolerance based autonomous robot navigation is the sensor differences even in the same robot. Even the physical sensors and actuators from the same batch may perform differently as they may be slightly different in their electronics or mechanics. This should also be taken into account in the fault-tolerant control design.

One approach to fault tolerance in the evolutionary robotics is morphology evolution [Mautner and Belew, 1999a; 1999b]. Except for the robotic controller evolution, it is also possible to evolve robots whose size and morphology are under the control of an evolutionary process. Sometimes, morphology can play an essential role since certain tasks may become much easier to perform for robots with certain particular shapes. Mostly importantly, morphology evolution provides the robot with the self-repairing capability such that the remaining components can be reconstructed via the evolutionary process. Thus the robotic system becomes more robust and reliable in various harsh conditions.

1.5.4 *Hardware Evolution*

The emerging application of artificial evolution to autonomous mobile robots enables robots to adapt their behaviors to changes of the environments in a continuous and autonomous fashion. For example, personal robots used for entertainment, education, and assistance interact with humans and service robots are being made to carry out work in unstructured environments [Keramas, 2000]. Meanwhile, Evolvable Hardware (EHW) has attracted much attention [Yao and Higuchi, 1999], which is a set of integrated circuits capable of adapting their hardware autonomously and in real time with a changing environment. Consequently, we have seen the fusion of these two technologies in these years. Artificial evolution can operate on reconfigurable electronic circuits to produce efficient and powerful control systems for autonomous mobile robots which are able to make their own decisions in complex and dynamic environment [Tan, et al., 2004]. Reconfigurable electronic circuit such as Field Programmable Gate Array (FPGA) can be viewed as a complex multi-input multi-output digital device, which is made up of a large number of Boolean functions. The input variables defined by the sensors on vehicle are fed into the FPGA and the output variables from the device are encoded into the different motor speeds. The multi-input multi-output function implemented on the FPGA should provide the desired reactive behaviors in response to the new changes in the environment.

Therefore, the evolution task can be clearly viewed as a searching process for the best system configuration whenever the environment changes. Intrinsic evolvable hardware best suits this application. Thompson used a reconfigurable hardware device (FPGA) to evolve a binary frequency discriminator that classifies its input signal into one of two frequencies [Thompson, Layzell, and Zebulum, 1999]. His search was GA-based and obtained its fitness measures in real time from the FPGA. Thompson identified the phenomenon of evolution exploiting the device's analog nature in finding solution circuits. The tone-discriminator experiment of Thompson clearly demonstrated intrinsic evolution's ability to explore rich structures and dynamical behaviors that are obviously radically different to those produced by conventional design, but yet which achieve the desired behavior perfectly. Evolving physical hardware directly instead of control systems simulated in software results in both higher speed and richer dynamics.

1.5.5 *On-Line Evolution*

One of the most urgent concerns is how evolved controllers can be efficiently evaluated. If they are tested using physical robots in the real world, then this should be conducted in real time. However, the evolution of complex behavior will take a prohibitively long time to evolve the desired behavior. If controllers are tested using simulation, then the amount of modeling needed to ensure that evolved controllers work on real robots may make the simulation too complex and computationally expensive. As a result, the potential speed advantage over the real-world evaluation is lost.

Up until now, several possible solutions have been proposed to solve the current problems in the evolution of adaptation tasks. For instance, Jakobi [1997] proposed the methodology based on minimal simulations which has been corroborated by a variety of experiments. The reliable transition from simulation to reality without performance loss is turned out to be feasible by modeling only a small subset of the robot/environmental properties [Jakobi, 1998; Jakobi and Quinn, 1998; Husbands, 1998; Smith, 1998]. However, as discussed previously, future work should pay more attention to the on-line learning schemes since they are more promising and useful in real-world applications in complex and dynamic environments to handle uncertainties. In the long sight, this might bring great advantages by accomplishing missions beyond human abilities. Grefenstette [1996] proposed that while operating in a real environment, an agent should maintain a simulated environmental model updated continuously whenever new features occur from the environment and the agent itself such that the learning continues throughout the operational phase. The work of Thompson [1995a] and Keymeulen et al. [1999] has already shown that the reliable transfer of behavior from simulation to reality is feasible without loss of robotic behaviors, though some issues still remain open. These ideas would definitely open up new opportunities in on-line robot evolution. Keymeulen et al. [1999] evaluated two approaches to the implementation of EHW to a robot navigation system: off-line model-free and on-line model-based evolution.

1.5.6 *Ubiquitous and Collective Robots*

In the recent decades, the fusion of information processing with physical processes has significantly changed the physical world around us and it also represents one of the most important opportunities for ER. The wide use of novel technologies such as System On a Chip (SOC), smart Microelectromechanical Systems (MEMS) transducers, Commercial-Off-The-Shelf (COTS) components, Internet connectivity, and high-dependability systems have brought great revolutions to the traditional real-time control systems [Kopetz, 2000; Prahlad, et al., 2002]. For example, inexpensive MEMS-based sensors and actuators have made it fairly feasible to integrate the physical world and information processing systems in various domains. Sensor networks are one of the most crucial applications of such embedded and distributed systems. Since wireless communication is used to reduce uncertainty in robotics [Sukhatme and Mataric, 2000] and ER is an effective way to deal with uncertainties, the fusion of these two innovative technologies is a natural and intuitive solution to handling uncertainty more efficiently. By introducing ER into distributed and embedded systems, there are many promising research fields ahead such as distributed and embedded robotic system based on wireless networks [Sukhatme and Mataric, 2000]. Here the robotics is used for network communication through physical mobility [Sukhatme and Mataric, 2000]. It can also be viewed as an application of ubiquitous computing (ubicomp) concept [Weiser, 1993] in robotics and the integration of distributed ER and embedded real-time system would incur new compelling and interesting topics to be resolved for this research community.

The distributed ER is also closely related to collective robotics. The term collective behavior generically refers to any behavior of a system having more than one robot. The use of multiple robots is often suggested to have many advantages over single-robot systems. Cooperating robots have the potential to accomplish the desired tasks more efficiently than a single robot [Xu, et al., 2002]. Furthermore, using several low-cost robots introduces redundancy and therefore is more fault-tolerant than having only one powerful and expensive robot. Therefore, collective behaviors offer the possibility of enhanced task performance, increased task reliability, and decreased cost over traditional robotic systems. In developing a multi-robot system, one of the primary concerns is how to enable individual robots to automatically generate task-handling behaviors adaptive to the dynamic changes in their task environments. The interactions among robots implicate development since they are not obvious in a particular robot but only emerge in an operating group.

1.6   Summary

Free-navigating mobile robotic systems can be used to perform service tasks for a variety of applications such as transport, surveillance, fire fighting, and so forth. For such robotic application systems, it is crucial to derive simple robotic behaviors that guarantee robust operation despite of the limited knowledge prior to system execution, e.g., designing an anti-collision behavior that is effective in the presence of unknown obstacle shapes. In recent years, autonomous mobile service robots have been introduced into various non-industrial application domains including entertainment, security, surveillance, and healthcare. They can carry out the cumbersome work due to their high availability, fast task execution, and cost-effectiveness. An autonomous mobile robot is essentially a computational system that acquires and analyzes sensory data or exterior stimulus and executes behaviors that may affect the external environment. It decides independently how to associate sensory data with its behaviors to achieve certain objectives. Such an autonomous system is able to handle uncertain problems as well as dynamically changing situations. Evolutionary robotics turns out to be an effective approach to realizing this purpose. The main theme of this chapter is to look into the applications of evolutionary approach in autonomous robotics. A general survey is reported regarding the effectiveness of a variety of artificial evolution based strategies in robotics. The open issues and future prospects in evolutionary robotics are also discussed. Some questions need to be answered if evolutionary robotics is to progress beyond the proof-of-concept stage. Furthermore, future prospects including SAGA, combination of learning and evolution, inherent fault tolerance, hardware evolution, on-line evolution, and ubiquitous and collective robots are suggested. With the infusion of more and more innovative ideas and technologies, we can see the bright future in this field, though there is a long way to go.

Bibliography

Aravinthan, A. and Nanayakkara, T. (2004), *Implementing Behavior Based Control in an Autonomous Mine Detecting Robot (AMDR)*, online available at: http://www.elect.mrt.ac.lk/B4_eru04.pdf.

Arsene, C. T. C., and Zalzala, A. M. S. (1999). Control of autonomous robots using fuzzy logic controllers tuned by genetic algorithms, *Proceedings of the International Congress on Evolutionary Computation 1999*, pp. 428–435.

Barfoot, T. D., and D'Eleuterio, G. M. T. (1999). An evolutionary approach to multiagent heap formation, *Proceedings of the International Congress on Evolutionary Computation 1999*, pp. 420–427.

Beer, R. D., and Gallagher, J. C. (1992). Evolving dynamic neural networks for adaptive behavior. *Adaptive Behavior*, 1, pp. 91–122.

Beetz, M. (2002). *Plan-Based Control of Robotic Agents, Lecture Notes in Artificial Intelligence 2254*, Springer-Verlag.

Beetz, M., Hertzberg, J., Ghallab, M., and Pollack, M. E. (2002). *Advances in Plan-Based Control of Robotic Agents, Lecture Notes in Artificial Intelligence 2466*, Springer-Verlag.

Berlanga, A., Isasi, P., Sanchis, A., et al. (1999). Neural networks robot controller trained with evolution strategies, *Proceedings of the International Congress on Evolutionary Computation*, pp. 413–419.

Brooks, R. (1992). Artificial life and real robots. F. J. Varela and P. Bourgine (Eds.), *Proceedings of the First European Conference on Artificial Life*, pp. 3–10. Cambridge MA: MIT Press/Bradford Books.

Brooks, R. (2002). Humanoid robots, *Communications of the ACM*, March, vol. 45, no. 3, pp. 33–38.

Brooks, R. A. (2001). Steps towards living machines, T. Gomi, (ed.): ER2001, *LNCS 2217*, pp. 72–93.

Chaiyaratana, N. and Zalzala, A. M. S. (1999). Hybridisation of neural networks and genetic algorithms for time-optimal control, *Proceedings of the International Congress on Evolutionary Computation 1999*, pp. 389–396.

Chavas, J., Corne, C., Horvai, P., et al. (1998). Incremental evolution of neural controllers for robust obstacle-avoidance in Khepera, *Proceedings of the First European Workshop on Evolutionary Robotics (EvoRobot98)*, France, pp. 227–247.

Chocron, O. and Bidaud, P. (1999). Evolving walking robots for global task based design, *Proceedings of the International Congress on Evolutionary Computation*, pp. 405–412.

Cliff, D., Harvey, I., and Husbands, P. (1993). Explorations in evolutionary robotics, *Adaptive Behavior*, 2, pp. 73–110.

Connell, J. H. (1992). SSS: A hybrid architecture applied to robot navigation, *Proceedings of the 1992 IEEE Conference on Robotics and Automation (ICRA92)*, pp. 2719–2724.

Dittrich, P., Burgel, A. and Banzhaf., W. (1998). Learning to move a robot with random morphology, *Proceedings of the First European Workshop on Evolutionary Robotics (EvoRobot 98)*, France, pp. 165–178.

Dorigo, M., and Colombetti, M. (1994). Robot shaping : Developing autonomous agent through learning. *Artificial Intelligence*, 71, pp. 321–370.

Dorigo, M., and Schnepf, U. (1993). Genetics-based machine learning and behavior-based robotics: A new synthesis. *IEEE Transactions on Systems, Man, Cybernetics*, 23, pp. 141–154.

Driankov, D., and Saffiotti, A. (Eds). (2001). *Fuzzy Logic Techniques For Autonomous Vehicle Navigation*, Springer-Verlag.

Driscoll, J. A., and Peters II, R. A. (2000). A development environment for evolutionary robotics, *Proceedings of the IEEE International Conference on Systems, Man, and Cybernetics*, pp. 3841–3845.

Earon, E. J. P., barfoot, T. D., and D'Eleuterio, G. M. T. (2000). From the sea to the sidewalk: the evolution of hexapod walking gaits by a genetic algorithm, *Evolvable Systems: From Biology to Hardware, Lecture Notes in Computer Science 1801 (Proc. of ICES2000)*, pp. 51–60, Springer-Verlag.

Floreano, D. (1997). Evolutionary robotics, an email interview conducted by M. Kelly and G. Robertson, online document is available at: http://dmtwww.epfl.ch/isr/east/team/floreano/evolrob.interview.html

Floreano, D., and Mattiussi, C. (2001). Evolution of spiking neural controllers for autonomous vision-based robots, T. Gomi, (ed.): ER2001, *LNCS 2217*, pp. 38–61.

Floreano, D., and Mondada, F. (1996). Evolution of homing navigation in a real mobile robot. *IEEE Transactions on Systems, Man and Cybernetics - Part B*, 26 (3).

Floreano, D., and Urzelai, J. (2000), Evolutionary robots with on-line self-organization and behavioral fitness, *Neural Networks*, 13, pp. 431–443.

Fogel, D. B. (2000a). What is evolutionary computation? *IEEE Spectrum*, pp. 26–32.

Fogel, D. B. (2000b). *Evolutionary Computation: Toward a New Philosophy of Machine Intelligence*, Second Edition, IEEE Press.

N. J. Fukuda, T., Mizoguchi, H., Sekiyama, K., and Arai, F. (1999). Group behavior control of MARS (Micro Autonomous Robotic System), *Proceedings of the IEEE International Conference on Robotics and Automation*, Detroit, pp. 1550–1555.

Full, R. J. (2001). Using biological inspiration to build artificial life that locomotes, T. Gomi, (ed.): ER2001, *LNCS 2217*, pp. 110–120.

Goldberg, D. E. (1989). *Genetic Algorithms In Search, Optimization and Machine Learning*. Reading, MA: Addison Wesley.

Gomi, T., and Griffith, A. (1996). Evolutionary robotics - an overview, *Proceedings of IEEE International Conference on Evolutionary Computation*, pp. 40–49.

Gomi, T., and Ide, K. (1998). Evolution of gaits of a legged robot, *Proceedings of the IEEE World Congress on Computational Intelligence*, pp. 159 –164.

Grefenstette, J. J. (1989). Incremental learning of control strategies with genetic algorithms. *Proceedings of the Sixth International Workshop on Machine Learning*, pp. 340–344.

Morgan Kaufmann. Grefenstette, J. J. (1996). Genetic learning for adaptation in autonomous robots, Robotics and manufacturing: Recent Trends in Research and Applications. vol. 6, ASME Press, New York.

Grefenstette, J. J., and Schultz, A. (1994). An evolutionary approach to learning in robots. *In Proceedings of the Machine Learning Workshop on Robot Learning, 11th International Conference on Machine Learning*, Morgan Kaufmann.

Gruau, F., and Quatramaran, K. (1997). Cellular encoding for interactive evolutionary robotics, *Proceedings of the Fourth European Conference on Artificial Life*. The MIT Press/Bradford Books.

Hagras, H., Callaghan, V., and Colley, M. (2001). Outdoor mobile robot learning and adaptation, *IEEE Robotics and Automation Magazine*, Sep., 2001, pp. 53–69.

Harvey, I. (1992). Species Adaptation Genetic Algorithms: A basis for a continuing SAGA, *Toward a Practice of Autonomous Systems: Proceedings of the First European Conference on Artificial Life*, F. J. Varela and P. Bourgine (eds.), MIT Press/Bradford Books, Cambridge, MA, pp. 346–354.

Harvey, I. (2001). Artificial evolution: a continuing SAGA, T. Gomi, (ed.): ER2001, *LNCS 2217*, pp. 94–109.

Higuchi, T., Iba, H., and Manderick, B. (1994). Applying evolvable hardware to autonomous agents, *Parallel Problem Solving From Nature (PPSNIII)*, pp. 524–533.

Hoffman, F., Koo, T. J. and Shakernia, O. (1998), Evolutionary design of a helicopter autopilot, *Proceedings of the 3rd On-line World Conference on Soft Computing*, Cranfield, UK.

Holland, J. H. (1975). *Adaptation In Natural And Artificial Systems*. Ann Arbor: The University of Michigan Press.

Holland, O. (2001). From the imitation of life to machine consciousness, T. Gomi, (ed.): ER2001, *LNCS 2217*, pp. 1–37.

Hopgood, A. A. (2003), Artificial intelligence: hype or reality? *IEEE Computer*, May, pp. 24–28.

Hornby, G. S., Takamura, S., Hanagata, O., et al. (2000), Evolution of controllers from a high-level simulator to a high DOF robot, *Evolvable Systems: From Biology to Hardware, Lecture Notes in Computer Science 1801 (Proc. of ICES2000)*, Springer-Verlag.

Hoshino, T., Mitsumoto, D., and Nagono, T. (1998). Fractal fitness landscape and loss of robustness in evolutionary robot navigation, *Autonomous Robots 5*, pp. 199–213.

Huang, S. H. (2002). Artificial neural networks and its manufacturing application: Part I, Online slides are available at: www.min.uc.edu/icams/resources/ANN/ANNManuI.ppt.

Husbands, P. (1998). Evolving robot behaviors with diffusing gas networks, *Proceedings of the First European Workshop on Evolutionary Robotics 98 (EvoRobot98)*, France, pp. 71–86.

Ishiguro, A., Tokura, S., and Kondo, T., et al. (1999). Reduction of the gap between simulated and real environments in evolutionary robotics: a dynamically-rearranging neural network approach, *Proceedings of the IEEE International Conference on Systems, Man, and Cybernetics*, pp. 239–244.

Jakobi, N. (1997). Half-baked, ad-hoc, and noisy: minimal simulations for evolutionary robotics, In Husbands, P., and Harvey, I., (Eds.) *Proc., Forth European Conference on Artificial Life*, MIT Press.

Jakobi, N. (1998). Running across the reality gap: Octopod locomotion evolved in a minimal simulation, *Proceedings of the First European Workshop on Evolutionary Robotics 98 (EvoRobot98)*, France, pp. 39–58.

Jakobi, N., and Quinn, M. (1998). Some problems (and a few solutions) for open-ended evolutionary robotics, *Proceedings of the First European Workshop on Evolutionary Robotics 98 (EvoRobot98)*, France, pp. 108–122.

Katagami, D., and Yamada, S. (2000). Interactive classifier system for real robot learning, *Proceedings of the 2000 IEEE International Workshop on Robot and Human Interactive Communication*, Japan, pp. 258–263.

Keramas, J. G. (2000). How will a robot change your life? *IEEE Robotics and Automation Magazine*, March, pp. 57–62.

Keymeulen, D., Iwata, M., Kuniyoshi, Y., et al. (1999). Online evolution for a self-adapting robotic navigation system using evolvable hardware, *Artificial Life Journal*, vol. 4, no. 4, pp. 359–393.

Khatib, O., Brock, O., Chang, K.-S., et al. (2002). Robotics and interactive simulation. *Communications of the ACM*, March, vol. 45, no. 3, pp. 46–51.

Kikuchi, K. Hara, F., and Kobayashi, H. (2001). Characteristics of function emergence in evolutionary robotic systems - dependency on environment and task, *Proceedings of the 2001 IEEE/RSJ International Conference on Intelligent Robots and Systems*, Hawaii, USA, pp. 2288–2293.

Kopetz, H. (2000). Software engineering for real-time: a roadmap. *Proceedings of the 22nd International Conference on Future of Software Engineering (FoSE) at ICSE 2000*, Limerick, Ireland.

Koza, J. R. (1994). *Genetic Programming II*, The MIT Press, Cambridge, Mass., USA.

Kubota, N., Morioka, T., Kojimi, F., et al. (1999). Perception-based genetic algorithm for a mobile robot with fuzzy controllers, *Proceedings of the International Congress on Evolutionary Computation 1999*, pp. 397–404.

Laue, T. and Rfer, T. (2004). A behavior architecture for autonomous mobile robots based on potential fields. *Proceedings of the 8th International Work-*

shop on RoboCup 2004 (Robot World Cup Soccer Games and Conferences), Lecture Notes in Artificial Intelligence. Springer, im Erscheinen.

Lee, W.-P., and Lai, S.-F. (1999). An example-based approach for evolving robot controllers, *IEEE Proceedings of the International Conference on Systems, Man, and Cybernetics*, pp. 618 –623.

Liu, J., Pok, C. K., and Keung, H. K. (1999). Learning coordinated maneuvers in complex environments: a sumo experiment, *Proceedings of the International Congress on Evolutionary Computation*, pp. 343–349.

Lund, H. H., and Hallam, J. (1997). Evolving sufficient robot controller, *Proceedings of the IEEE International Conference on Evolutionary Computation*, pp. 495–499.

Lund, H. H., and Miglino, O. (1996). From simulated to real robots, *Proceedings of IEEE International Conference on Evolutionary Computation*, pp. 362–365.

Lund, H. H., Bjerre, C., and Nielsen, J. H., et al. (1999). Adaptive LEGO robots: a robot = human view on robotics, *Proceedings of the IEEE International Conference on Systems, Man, and Cybernetics*, pp. 1017–1023.

Lund, H. H., Miglino, O., and Pagliarini, L., et al. (1998). Evolutionary robotics - a children's game, *Proceedings of the IEEE World Congress on Computational Intelligence*, pp. 154 –158.

Mautner C. and Belew, R. K. (1999a). Coupling Morphology and Control in a Simulated Robot, *Proceedings of the Genetic and Evolutionary Computation Conference*, vol. 2, Orlando, USA, pp. 1350–1357.

Mautner, C. and Belew, R. (1999b). Evolving Robot Morphology and Control, In: Sugisaka, M. (ed.), *Proceedings of Artificial Life and Robotics 1999 (AROB99)*, Oita.

Meeden, L. A., and Kumar, D. (1998). Trends in evolutionary robotics, *Soft Computing for Intelligent Robotics Systems*, edited by L. C. Jain and T. Fukuda, Physica-Verlag, NY, pp. 215–233.

Meyer, J.-A. (1998). Evolutionary approaches to neural control in mobile robots, *Proceedings of the IEEE International Conference on Systems, Man, and Cybernetics*, pp. 2418 –2423.

Meyer, J.-A., Husbands, P., and Harvey, I. (1998). Evolutionary robotics: A survey of applications and problems, *Proceedings of the First European Workshop on Evolutionary Robotics 98 (EvoRobot98)*, France, pp. 1–21.

Mitchell, T. M. (1997), *Machine Learning*, The McGraw-Hill Companies. Inc.

Mondada, F., and legon, S. (2001). Interactions between art and mobile robotic system engineering, T. Gomi, (ed.): ER2001, *LNCS 2217*, pp. 121–137.

Murase, K., Wakida, T., Odagiri, R., et al. (1998). Back-propagation learning of autonomous behavior: a mobile robot Khepera took a lesson from the future consequences, *Evolvable Systems: From Biology to Hardware, Lecture Notes in Computer Science 1478 (Proc. of ICES1998)*, Springer-Verlag.

Nicolescu, M. N. and Mataric, M. J. (2002). A hierarchical architecture for behavior-based robots, *Proceedings of the First International Joint Conference on Autonomous Agents and Multi-Agent Systems*, Bologna, Italy.

Nolfi, S. (1998a). Adaptation as a more powerful tool than decomposition and integration: experimental evidences from evolutionary robotics, *Proceedings of the IEEE World Congress on Computational Intelligence*, pp. 141 –146.

Nolfi, S. (1998b). Evolutionary robotics: exploiting the full power of self-organizing, *Self-Learning Robots II: Bio-robotics (Digest No. 1998/248)*, Page(s): 3/1 –3/7.

Nolfi, S. Floreano, D., and Miglino, O., et al. (1994). How to evolve autonomous robots: different approaches in evolutionary robotics, In R. A. Brooks and P. Maes (Eds.), *Proceedings of the IV International Workshop on Artificial Life*, Cambridge, MA: MIT Press.

Nolfi, S., and Floreano, D. (2000). *Evolutionary Robotics: The Biology, Intelligence, and Technology of Self-Organizing Machines*, MA: MIT Press.

Nordin, P. and Banzhaf, W. (1997). Real-time control of a Khepera robot using genetic programming, *Control and Cybernetics*, 26 (3), pp. 533–561.

Nordin, P., Banzhaf, W., and Brameier, M. (1998). Evolution of a world model for a miniature robot using genetic programming, *Robotics and Autonomous Systems*, 25, pp. 105–116.

Paraskevopoulos, P. N. (1996). *Digital Control Systems*, Prentice Hall.

Parisi, N. J. D., Nolfi, S., and Cecconi, F. (1992). Leaning, behavior and evolution. *Proceedings of the First European Conference on Artificial Life*, pp. 207–216. Cambridge, MA: MIT Press/Bradford Books.

Parker, G. B. (1999). The co-evolution of model parameters and control programs in evolutionary robotics, *Proceedings of IEEE International Symposium on Computational Intelligence in Robotics and Automation*, pp. 162 –167.

Pollack, J. B., Lipson, H., Ficci, S., et al. (2000). Evolutionary techniques in physical robotics, *Evolvable Systems: From Biology to Hardware, Lecture Notes in Computer Science 1801 (Proc. of ICES2000)*, pp. 175–186, Springer-Verlag.

Pollack, J. B., Lipson, H., Funes, P., and Hornby, G. (2001). First three generations of evolved robots, T. Gomi, (ed.): ER2001, *LNCS 2217*, pp. 62–71.

Prahlad, V., Ko, C. C., Chen, B. M., et al. (2002), Development of a Web-based mobile robot control experiment, *Proc. of 2002 FIRA Robot World Congress*, Korea, pp. 488 – 493.

Pratihar, D. K., Deb, K., and Ghosh, A. (1999). Fuzzy-genetic algorithms and mobile robot navigation among static obstacles, *Proceedings of the International Congress on Evolutionary Computation 1999*, pp. 327–334.

Rechenberg, I. (1973). *Evolutionstrategie: optimierung technischer systeme nach pronzipien der biologischen evolution*. Stuttgart: Friedrich Fromann Verlag.

Reil, T., and Husbands, P. (2002). Evolution of central pattern generators for bipedal walking in a real-time physics environment, *IEEE Transactions on Evolutionary Computation*, vol. 6, no. 2, April, pp. 159–168.

Revello, T. E., and McCartney, R. (2000). A cost term in an evolutionary robotics fitness function, *Proceedings of the 2000 Congress on Evolutionary Computation*, pp. 125 –132.

Reynolds, C. W. (1994). An evolved, vision-based model of obstacle avoidance behavior. *Artificial Life III*, pp. 327–346. Reading, MA: Addison-Wesley.

Rus, D., Butler, Z., Kotay, K., and Vona, M. (2002). Self-reconfiguring robots.

Communications of the ACM, March, vol. 45, no. 3, pp. 39–45.

Saffiotti, A. (1997). The uses of fuzzy logic in autonomous robot navigation, *Soft Computing*, Springer-Verlag, pp. 180–197.

Schaefer, C. G. (1999). Morphogenesis of path plan sequences through genetic synthesis of L-system productions, *Proceedings of the International Congress on Evolutionary Computation*, pp. 358–365.

Schraft, R. D. and Schmierer, G. (2000), *Service robots*, A K Peters, Ltd., MA.

Secchi, H., Mut, V., Carelli, R. Schneebeli, H., Sarcinelli, M. and Bastos, T. F. (1999), A hybrid control architecture for mobile robots: classic control, behavior based control, and petri nets. http://www.inaut.unsj.edu.ar/Files/Se1432_99.pdf

Seth, A. K. (1998). Noise and the pursuit of complexity: A study in evolutionary robotics, *Proceedings of the First European Workshop on Evolutionary Robotics 98 (EvoRobot98)*, France, pp. 123–136.

Smith, T. M. C. (1998). Blurred vision: Simulation-reality transfer of a visually guided robot, *Proceedings of the First European Workshop on Evolutionary Robotics 98 (EvoRobot98)*, France, pp. 152–164.

Sridhar, M. and Connell, J. (1992). Automatic programming of behavior-based robots using reinforcement learning, *Artificial Intelligence*, vol. 55, nos. 2–3, pp. 311–365. http://robotics.jpl.nasa.gov/people/ericb/ieee_aero00_h.pdf

Staugaard, Jr., A. C. (1987). *Robotics and AI: An Introduction to Applied Machine Intelligence*, Prentice-Hall, Englewood Cliffs, N. J.

Steels, L. (1995). Discovering the competitors. *Adaptive Behaviors 4*, pp. 173–199.

Sukhatme, G. S., and Mataric, M. J. (2000). Embedding robots into the Internet. *Communications of the ACM*, May, vol. 43, no. 5, pp. 67–73.

Sukhatme, G. S., and Mataric, M. J. (2002). Robots: intelligence, versatility, adaptivity. *Communications of the ACM*, March, vol. 45, no. 3, pp. 30–32.

Sutton, R.S. and Barto, A.G. (1998), *Reinforcement Learning: An Introduction*, MIT Press.

Sutton, R.S. (1988), Learning to predict by the methods of temporal difference, *Machine Learning 3*, pp. 9–44.

Tan, K. C., Lee, T. H., Khoo, D., and Khor, E. F. (2001a). A multi-objective evolutionary algorithm toolbox for computer-aided multi-objective optimization, *IEEE Transactions on Systems, Man and Cybernetics: Part B (Cybernetics)*, vol. 31, no. 4, pp. 537–556.

Tan, K. C., Lee, T. H., and Khor, E. F. (2001b). Evolutionary algorithm with dynamic population size and local exploration for multiobjective optimization, *IEEE Transactions on Evolutionary Computation*, vol. 5, no. 6, pp. 565–588.

Tan, K. C., Chew, C. M., Tan, K. K., and Wang, L. F., et al. (2002a). Autonomous robotic navigation via intrinsic evolution, *IEEE Proceedings of the 2002 World Congress on Evolutionary Computation*, Hawaii, USA, pp. 1272–1277.

Tan, K. C., Tan, K. K., Lee, T. H., et al. (2002b). Autonomous robot navigation based on fuzzy sensor fusion and reinforcement learning, *Proceedings of the 17th IEEE International Symposium on Intelligent Control (ISIC'02)*,

Vancouver, British Columbia, Canada, pp. 182–187.

Tan, K. C., Wang, L. F. Lee, T. H., and Prahlad, V. (2004). Evolvable Hardware in Evolutionary Robotics, *Autonomous Robots*, vol. 16, issue 1, pp. 5–21.

Taubes, G. (2000). Biologists and engineers create a new generation of robots that imitate life, *Science*, vol. 288, no. 5463, issue of 7, Apr 2000, pp. 80–83.

Thompson, A. (1995). Evolving electronic robot controllers that exploit hardware resources. *Proc. of the 3rd European Conf. on Artificial Life (ECAL95)*, pp. 640–656, Springer-Verlag.

Thompson, A., Layzell, P., and Zebulum, R. S. (1999). Explorations in design space: unconventional electronics design through artificial evolution, *IEEE Transactions on Evolutionary Computation*, vol. 3, no. 3, pp. 167–196.

Thrun, S. (2002). Probabilistic robotics. *Communications of the ACM*, March, vol. 45, no. 3, pp. 52–57.

Veloso, M. M. (2002). Entertainment robotics, *Communications of the ACM*, March, vol. 45, no. 3, pp. 59–63.

Walker, J. F., and Oliver, J. H. (1997). A survey of artificial life and evolutionary robotics, on-line document is available at: http://citeseer.nj.nec.com/walker97survey.html.

Wang, L. F. (2002), Computational intelligence in autonomous mobile robotics - a review, *IEEE Proceedings of the Thirteenth International Symposium on Micromechatronics and Human Science*, Nagoya, Japan, Oct., pp. 227–235.

Watson, R. A., Ficici, S. G., and Pollack, J. B. (1999). Embodied Evolution: Embodying an evolutionary algorithm in a population of robots, *Proceedings of the 1999 Congress on Evolutionary Computation*.

Weiser, M. (1993). Some computer science issues in ubiquitous computing, *Communications of the ACM*, 36(7), July, pp. 75–84.

Xiao, P., Prahlad, V., Lee, T. H., and Liu, X. (2002), Mobile robot obstacle avoidance: DNA coded GA for FLC optimization, *Proc. of 2002 FIRA Robot World Congress*, Korea, pp. 553 – 558.

Xu, L., Tan, K. C., Prahlad, V., and Lee, T. H. (2002). Multi-agents competition and cooperation using fuzzy neural systems, *Proceedings of the 4th Asian Control Conference*, Singapore, pp. 1326–1331.

Yamada, S. (1998). Learning behaviors for environmental modeling by genetic algorithm, *Proceedings of the First European Workshop on Evolutionary Robotics 98 (EvoRobot98)*, France, pp. 179–191.

Yao, X., and Higuchi, T. (1999). Promises and challenges of evolvable hardware, *IEEE Trans. on Systems, Man, and Cybernetics- Part C: Applications and Reviews*, vol. 29, no. 1, pp. 87–97.

Chapter 2

Evolvable Hardware in Evolutionary Robotics*

In the recent decades the research of Evolutionary Robotics (ER) has been developed rapidly, which is primarily concerned with the use of evolutionary computing techniques for automatic design of adaptive robots. Meanwhile, much attention has been paid to a new set of integrated circuits named Evolvable Hardware (EHW), which is capable of reconfiguring their architectures using artificial evolution techniques unlimited times. This chapter surveys the application of evolvable hardware in evolutionary robotics, which is an emerging research field concerning the development of evolvable robot controller at the hardware level to adapt dynamic changes in the environments. The context of evolvable hardware and evolutionary robotics is reviewed respectively, and a few representative experiments in the field of robotic hardware evolution are presented. As an alternative to conventional robotic controller designs, the potentialities and limitations of the EHW-based robotic system are discussed and summarized.

2.1 Introduction

In the recent decades researchers have been working on the application of artificial evolution to autonomous mobile robots capable of adapting their behaviors to changes in the physical environments. For example, the robots in the next generation should be able to interact with humans and carry out work in unstructured environments (Keramas, 2000). As a result, an infant research field called Evolutionary Robotics (ER) has been developed rapidly, which is primarily concerned with the use of evolutionary computing techniques for automatic design of adaptive robots. Meanwhile, much

*With kind permission of Springer Science and Business Media.

attention has been paid to the development of Evolvable Hardware (EHW) (Sanchez et al., 1996; Kitano, 1996; Manderick and Higuchi, 1996; Higuchi et al., 1996b; Yao, 1999; Sipper and Ronald, 2000), which is a new set of integrated circuits capable of reconfiguring their architectures using artificial evolution techniques. Hardware evolution dispenses with conventional hardware designs in solving complex problems in a variety of application areas, ranging from pattern recognition (Iwata et al., 1996; Higuchi et al., 1997a) to autonomous robotics (Thompson, 1995a; Keymeulen et al., 1997; Haddow and Tufte, 1999).

This chapter mainly discusses the application of evolvable hardware in evolutionary robotics. The concept and classification of evolvable hardware are presented in Section 2.2. The context of evolutionary robotics and the current works in the area are reviewed in Section 2.3. In Section 2.4, the application of evolvable hardware in evolutionary robotics is reviewed via a few representative experiments in the field. As an alternative to conventional robotic controller designs, the potentialities and limitations of the EHW-based robotic systems are discussed and summarized in Section 2.5. Conclusions are drawn in Section 2.6.

2.2   Evolvable Hardware

This section reviews the body of much related literature regarding EHW, which could be useful for understanding the concept of EHW. In particular, some important features are discussed that can be used for the classification of EHW, such as artificial evolution, reconfigurable hardware device, extrinsic and intrinsic evolutions, etc.

2.2.1   *Basic Concept of EHW*

EHW is based on the idea of combining reconfigurable hardware devices with evolutionary algorithms (e.g., genetic algorithm (GA)) to execute reconfiguration automatically. Artificial evolution and reconfigurable hardware device are the two essential elements in EHW. The basic concept behind the combination of these two elements in EHW is to regard the configuration bits for reconfigurable hardware devices as chromosomes for genetic algorithms (Goldberg, 1989; Higuchi et al., 1999). Fig. 2.1 illustrates the general evolution process in an evolvable hardware [Keymeulen, et al., 1999]. Typically, an initial population of individuals is randomly generated. Using a user-defined fitness function (the desired hardware performance), the GA selects promising individuals in the population to reproduce offspring for the next generation based upon the Darwinian principle of survival-of-the-fittest (Goldberg, 1989; Tan et al., 2001). If the fitness function is properly designed for a specific task, then the GA can automatically find the best hardware configuration in terms of architecture bits to realize the desired task.

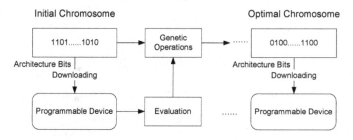

Fig. 2.1   The evolution mechanism of evolvable hardware using genetic algorithms.

2.2.2   *Classification of EHW*

EHW can be classified along the lines of artificial evolution, hardware device, evolution process, adaptation method, and application area as illustrated in Fig. 2.2 (Zebulum et al., 1996). In this subsection, the classification of EHW along these dimensions is discussed.

2.2.2.1   *Artificial evolution and hardware device*

The two essential elements in EHW are artificial evolution and reconfigurable hardware device:

A. Choice of EA: Genetic Algorithms (GAs) (Goldberg, 1989; Holland, 1975) are currently the most often used Evolutionary Algorithms (EAs) (Fogel, 2000) in EHW research, which are capable of seeking solutions effectively from a vast and complex search space. The GA is a stochastic search algorithm that has proved useful in finding the global optimum in both the static and dynamic environments. It works with a population of candidate solutions that are repeatedly subjected to selection pressure (survival-of-the-fittest) in the search for better solutions. The experiments of Thompson (1995a) demonstrated in particular that artificial evolution in the form of GA can serve as an effective training tool for hardware evolu-

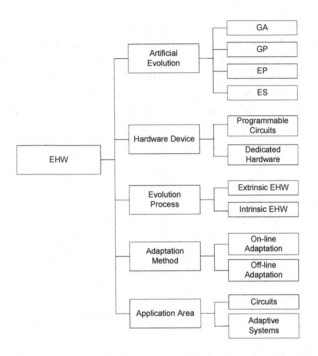

Fig. 2.2   Classification of the EHW.

tion. Thompson evolved discriminator circuits properly which are capable of classifying their inputs correctly when one of the two possible input frequencies is presented. Other non-conventional GAs such as Variable-length GA (VGA) (Iwata et al., 2000), Species Adaptation Genetic Algorithm (SAGA) (Harvey, 1992), and Production GA (PGA) (Mizohuchi et al, 1994) have also been successfully applied to specific evolution tasks.

Another field of study in EHW is the application of Genetic Programming (GP) (Koza, 1994) principles to high-level hardware design languages (Koza et al., 1996; Sakanashi et al., 1996; Bennett et al., 1996). The GP is also based upon the Darwinian idea of natural selection and genetic recombination, where individuals are often represented as tree-structures, i.e., in this context, the bit string comprising of an individual is interpreted as a sequence or tree of programming language. It does not need to know the operating environment and the designer only needs to devise the fitness evaluators to guide the evolution of programs towards the required task. The GP techniques have been employed in circuit descriptions using hardware description language called SDL (Structured function Description Language) (Koza et al. 1996; Sipper, 1996; Stauffer and Sipper, 1998). Since GP usually works with a large number of programs or individuals represented in tree-structures, there is a time penalty to be considered when it is applied to EHW applications (Lazarus and Hu, 2001).

Besides the above commonly used EAs, other EAs such as Evolutionary Strategy (ES) (Ros, 1997) and Evolutionary Programming (EP) (Michalewicz, 1994) have also been used in EHW research, and many concerns and discussions are shared among these paradigms.

B. Choice of reconfigurable hardware device: There are many reconfigurable hardware devices in the market, and the prevalent technology used in the field of EHW today is Field Programmable Gate Array (FPGA) (Shirasuchi, 1996; Xilinx Inc., 1996). The FPGA has an array of logic cells placed in an infrastructure of interconnections, which can be configured via a string of bits called architecture bits [Keymeulen, et al., 1999]. Although FPGA is a popular choice in EHW due to its flexibility and rapid reprogrammability, dedicated architecture devices are also available for specific applications. Several research groups are working on alternative hardware platforms for the EHW (Hamilton et al., 1998; Higuchi and Kajihara, 1999; Higuchi et al., 1999; Layzell, 1999; Sakanashi et al., 1999; Haddow and Tufte, 2000).

2.2.2.2 *Evolution process*

In the previous subsection, EHW was classified into different categories according to the different EAs and electronic circuits used. However, there are also two other dimensions worth consideration in investigating the EHW (Hirst, 1996; Yao, 1999; Yao and Higuchi, 1999). For instance, how the simulated evolution is implemented and what the simulated evolution is used for (Zebulum et al., 1996). This subsection considers the EHW along these two dimensions.

EHW is often implemented on Programmable Logic Devices (PLDs) such as FPGAs. The architecture of PLD and its function are determined by a set of architecture bits which can be reconfigured. In EHW, the simulated evolution is used to evolve a good set of architecture bits in order to solve a particular problem. According to de Garis (1996), EHW can be classified into the categories of extrinsic and intrinsic EHW. Extrinsic EHW simulates the evolution in software and downloads the best configuration to hardware in each generation, i.e., the hardware is reconfigured only once in each generation. Intrinsic EHW simulates the evolution directly in its hardware. Every chromosome is used to reconfigure the hardware and therefore the EHW is reconfigured that many times as the population size in each generation (Yao and Higuchi, 1999). Here the distinction between intrinsic and extrinsic evolution should be noted: it only concerns whether an EHW is reconfigured once or multiple times in each generation.

In extrinsic EHW, the evolution process can be easily carried out in a host computer through software simulation. However, it is often difficult to model even a small subset of physical properties of the hardware using conventional circuit simulation tools such as SPICE (Quarles et al., 1994). The more abstract a model is, the further it is away from the hardware medium. Intuitively, if all the detailed characteristics of a hardware could be simulated perfectly in extrinsic EHW (which is often very computationally expensive), then the two approaches are equivalent in the evolution process. However, in real-world applications it is not practical to model the complete hardware resources at the expense of computational efficiency. Much work on extrinsic EHW has been reported in literature including Higuchi et al. (1993), Hemmi et al. (1994), Koza et al. (1996), Lohn and Colombano (1998), etc.

The fact that physical modeling of the hardware medium is not necessary in intrinsic EHW offers two major advantages over extrinsic EHW, i.e., higher evaluation speed and richer continuous dynamics (Higuchi et al., 1996a; Thompson, 1996a, 1996b; Thompson et al., 1999). Thompson (1996a; 1996b) implemented an intrinsic EHW by evolving a circuit to discriminate between square waves of 1 kHz and 10 kHz presented at the input. The output is expected to rise to +5 V as soon as one of the frequencies is present and fall to 0 V for the other. The task is not easy because only a few components were provided and the circuit had no access to a clock. An optimal arrangement of 100 cells in the EHW chip needs to be derived so as to perform the task entirely on-chip. To configure a single cell, 18 configuration bits need to be set and these bits were directly encoded on a linear bit-string genotype. To evaluate the fitness of an individual, the 1800 bits of the genotype were used to configure the 10×10 corner of the FPGA one-by-one. Each chromosome was evaluated according to its performance in the real hardware. With a population of 50, the desired behavior was obtained in 3500 generations. In addition, the circuit's robustness and fault tolerance (Thompson, 1995b; 1996c; 1997a) were explored, which show that the approach of intrinsic EHW enables the evolution to fully exploit the circuit's characteristics.

Besides the methods of extrinsic and intrinsic hardware evolution, Stoica et al., (2000) introduced the concept of mixtrinsic EHW. In this approach, the evolution takes place with hybrid populations, in which some individuals are evaluated intrinsically and some extrinsically, within the same generation or in alternate generations.

2.2.2.3 *Adaptation methods*

The main attractiveness of EHW comes from its potential as an on-line adaptive hardware which is capable of changing its behavior to achieve better performance while executing in a real physical environment (Yao and Higuchi, 1999). Hence the EHW can also be classified along the approach of adaptation. In off-line EHW, the adaptation phase precedes the execution phase, i.e., the adaptation happens during the learning phase instead of the execution mode of the EHW. In this approach, the closeness of the simulated environment to the real-world has a major impact on the performance of the evolved hardware in the physical environment. As stated by Nolfi et al., (1994), one merit of using the off-line hardware evolution is that it allows a preliminary study of the evolution process prior to the real-time physical implementation.

Due to the trial-and-error search nature of GAs, individuals with poor performance in on-line EHW adaptation may cause severe damage to the EHW, the robot or the physical environment in which it is being evaluated if no preventive measure is employed (Yao and Higuchi, 1999). Moreover, the population-based evolutionary optimization in EHW at present cannot make the learning to be incremental and responsive. Besides the drawback of high computation expense due to its real-time interactions with the environments, on-line adaptation also faces the problem of evaluating the fitness function. For example, a mobile robot often needs to decide the motor speeds based on the sensory information in on-line robotic navigation. However, different moves will generate different sensory information in subsequent time steps, and this information is used to evaluate the fitness function for the corresponding move. Therefore on-line EHW adaptation is often difficult to be achieved in practice, except for a few examples (Goeke et al., 1997; Yao and Higuchi, 1999).

2.2.2.4 *Application areas*

According to the application domains as shown in Fig. 2.2, EHW could be applied to design circuits and adaptive systems (Yao and Higuchi, 1999).

There are two views for the definition of EHW along this line. One view regards EHW as the application of evolutionary techniques to circuit synthesis (de Garis, 1996). This definition describes the EHW as an alternative to conventional specification-based electronic circuit designs. Another view regards EHW as the hardware capable of on-line adaptation by reconfiguring its architectures dynamically and automatically, i.e., it specifies the EHW as an adaptive mechanism (Higuchi and Kajihara, 1999). The application of EHW in ER discussed in this chapter falls into the latter category.

2.2.3   Related Works

Instead of evolving hardware at the gate-level (Iba et al., 1997), Higuchi proposed the concept of functional level EHW for various applications, such as data compression, neural networks, ATM control, evolutionary robotics and pattern recognition (Higuchi et al., 1996a; 1997b). At the macro hardware level, Lipson and Pollack (2000) evolved locomotive systems in computational space and used rapid-prototyping technology to automatically produce multi-linked structures that only need motors to be snapped on by hand (Brooks, 2000; Forbes, 2000). Mondada and Floreano (1995) evolved robotic controllers for real Khepera robots using neural network based evolutionary control. Lewis et al., (1994) evolved the walking patterns for walking robots using a staged evolution approach. The latter three examples are elaborated in Section 2.4.

2.3   Evolutionary Robotics

Traditional manipulator-based robot controllers are often programmed in an explicit way since the controlled robots normally execute certain repetitive or fixed actions in a well-defined environment. However, autonomous robotic applications often require the mobile robots to be able to react quickly to unpredictable situations in a dynamic environment without any human intervention. For instance, personal and service robots are two important robotic applications for the next generation. In both cases, what matters most is the necessity to adapt to new and unpredictable situations, to react quickly, to display behavioral robustness, and to operate in close interaction with human beings (Floreano and Urzelai, 2000). For such mobile robotic applications, however, it is difficult to define the ever-changing unstructured environments via sensors in sufficient detail. Therefore traditional robotic controllers designed for factory automation are not suitable for these types of applications.

In the wake of the spread of Darwinian thinking, evolutionary computing techniques (Holland, 1975; Goldberg, 1989) have been widely applied to robotic control applications (Jakobi, 1997; Jakobi and Quinn, 1998; Husbands et al., 1994; Gomi and Griffith, 1996; Meyer et al., 1998). As an infant research field, evolutionary robotics is primarily concerned with the use of evolutionary computing techniques for automatic design of adaptive robots. It represents a powerful approach to enable robots to adapt to unpredictable environments. The robot controllers derived by such evolutionary techniques often provide a relatively fast adaptation time and carefree operations (Nolfi and Floreano, 2000).

Many approaches have been applied to the field of evolvable robotic controller designs (Chocron and Bidaud, 1999; Pollack et al., 2000). A number of researchers used artificial neural networks (ANNs) of some variety as the basic building blocks for robotic control systems, due to the smooth search space and the low primitives handling capability (Floreano and Mondada, 1994; Odagiri et al., 1998; Miglino et al., 1998). Other advantages of ANN include its learning and adaptation through efficient knowledge acquisition, domain free modeling, robustness to noise, fault tolerance and etc. (Huang, 2002). For instance, Floreano and Mondada (1994; 1996) presented an evolutionary system with discrete-time recurrent neural network to create an emergent homing behavior. Lund and Miglino (1996) evolved a neural network control system for Khepera robot using a simulator. Implemented upon a real Khepera robot, Kondo et al., (1999) incorporated the concept of dynamical rearrangement function of biological neural networks with the use of neuro-modulators to develop a robust robot controller. Although many examples of ANN in robotic applications have been reported, there are several drawbacks of ANN that need to be considered. For instance, the ANN is essentially a black-box system and the training associated is usually time-consuming. Furthermore, the learning algorithm of ANN may not always guarantee the convergence to an optimal solution (Huang, 2002).

As a commonly used EA, genetic algorithms have been applied to generate many desired robot behaviors in evolutionary robotics (Floreano and Mondada, 1996; Yamada, 1998; Barfoot and Eleuterio, 1999; Arsene and Zalzala, 1999; Kubota et al., 1999; Chaiyaratana and Zalzala 1999; Pratihar et al., 1999; Earon et al., 2000). The GA exhibits robust behavior in conditions where large numbers of constraints and/or training data are required (Walker and Oliver, 1997). It can also be applied to a variety of research communities due to its gene representation. However, the speed of convergence and the selection of cost function for specific robot behaviors are the main concerns of GAs in the application of ER (Lewis et al., 1992; Revello and McCartney, 2000).

Besides GA, genetic programming that evolves computer programs in the lisp or schema computer languages has also been used in the field of ER. Several experiments on evolving robot controllers using GP have been reported (Brooks, 1992; Koza, 1994; Reynolds, 1994; Dittrich et al., 1998; Liu et al., 1999; Schaefer, 1999). In particular, Nordin and Banzhaf (1997) used GP to control a robot trained in the real environments. The approach was later extended to allow learning from past experiences stored in the memory and excellent behaviors were emerged in the experiments (Nordin et al., 1998). As compared to GA, the GP has fewer applications in behavior-based robotics due to its tree-structures representation and low evolution speed. It is anticipated that GP will be applied to more practical ER applications as higher computational processing power becomes available (Lazarus and Hu, 2001).

Besides these commonly used methodologies in ER, several other approaches have also been proposed to work as an evolution mechanism for robotics. These approaches include classifier systems (Dorigo and Schnepf, 1993; Dorigo and Colombetti, 1994; Grefenstette and Schultz, 1994; Grefenstette, 1996), cellular encoding (Gruau and Quatramaran, 1997), evolutionary programming (Hasegawa et al., 1999), evolutionary strategy (Berlanga et al., 1999), embodied evolution (Watson et al., 1999), and etc.

Most of the evolutionary robotic design approaches discussed above are based on software simulations. For instance, evolutionary algorithms have been used to evolve artificial neural networks where software simulations cover all the evolution process. The application of evolvable hardware in evolutionary robotics makes it possible to apply evolutionary techniques to evolve patterns of behaviors for robots at the hardware level by fully exploiting the hardware resources and dynamics, which is discussed in the following section.

2.4   Evolvable Hardware in Evolutionary Robotics

As a thriving research field, EHW-based evolutionary robotics has been developed rapidly in the recent years (Floreano and Mondada, 1996; Naito et al., 1997; Yamamoto and Anzai, 1997; Yao and Higuchi, 1999). A few basic robotic behaviors have been derived using surprisingly tiny part of an EHW chip, which are comparable to or better than those achieved by conventional methods. In this section, several representative experiments of EHW in evolutionary robotics are described. The first three examples are concerned with robotic hardware evolution at the gate level and the following three examples are related to the macro hardware evolution for robot controller designs.

A. Thompson: Dynamic state machine based evolution

A particular architecture termed Dynamic State Machine (DSM) was proposed as a reconfigurable system for intrinsic hardware evolution (Thompson, 1995a, 1996a). In this architecture, all fitness evaluations associated with the individuals are instantiated as real configuration of the hardware implementation in DSM. The DSM architecture differs from finite state machine because a description of its state must include the temporal relationship between the asynchronous signals. The evolving DSM is intimately coupled with the real-time dynamics of its sensorimotor environment so that real-valued time can play an important role throughout the system. The evolving DSM can explore special-purpose tight sensorimotor coupling because the temporal signals can quickly flow through the system being influenced by, and in turn perturbing, the DSM on their way.

Thompson used the off-line adaptation approach to evolve a real hardware controller for Khepera robot in a real-world environment. The task is to induce the wall-avoidance behavior and to keep the robot moving within the center of an arena. As shown in Fig. 2.3, the robot's only sensors are a pair of time-of-flight sonars mounted on the robot, one pointing to the left and the other to the right. Other hardware devices necessary in conventional controller designs such as reflection timers and pulse-width modulators were not utilized in this experiment. The evolution was based on the configuration of RAM for controlling the DSM, and each individual represents a RAM configuration. Both the contents of RAM and clock frequency were under the evolutionary control, which endow the DSM with a new rich range of possible dynamical behavior so that it can be directly integrated with the sonars and motors. Using the "Genetic Latches" evolution strategy, it was able to choose which part of the circuit is synchronous to a clock, if any. The resulted circuit using a mixture of synchronous/asynchronous control was able to achieve the required behavior using only 32-bits of RAM and a few flip-flops to implement a pair of logic functions of four variables. Besides the above experiment, tolerance of the controller to faults in the memory array of the RAM chip was also implemented using the DSM architecture.

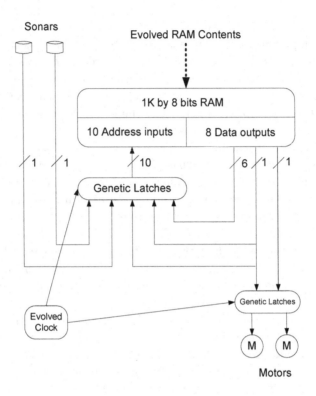

Fig. 2.3   The evolvable DSM robot controller.

In conventional robotic controller designs, a timer is necessary to measure the length of its output pulses for each of the sonars and thus the time-of-flight of the sound to indicate the range to the nearest object on that side of the robot. These timers provide binary-coded representations of the two flight times to a central controller. The central controller is a hardware implementation of the finite state machine responsible for computing a binary representation of the appropriate motor speed for each wheel and for controlling the wheel speeds via a pulse-width modulator. However, in the preceding robotic controller designs using intrinsic hardware evolution, the timers are not indispensable since the system clock is also under the evolutionary control, and the target task is accomplished in a more efficient fashion. The work can also be regarded as an investigation for chip independence-based hardware evolution. The experiment was later implemented on an XC6216 FPGA chip when the EHW was commercially available (Thompson, 1996d).

B. Haddow and Tufte: Complete hardware evolution

Based on the work of Thompson, Haddow and Tufte (1999) used a hardware evolution approach termed Complete Hardware Evolution (CHE) to evolve robotic controllers. In their approach as shown in Fig. 2.4, no work was executed in the host processor and all the evolution process was completed in a single FPGA chip consisting of a RAM for storing individuals of the population, the genetic pipeline for steering and evaluating the evolution of DSM, and the DSM itself. The genetic pipeline is responsible for the functions of genetic operations, fitness evaluation and sorting, which are implemented in separate modules in the pipeline. The individuals in the current generation are stored in the RAM and forwarded to the DSM one-by-one. The DSM controls the robot movement according to the current individual and provides the fitness value to the genetic pipeline. The new generation that is created after the genetic reproduction is also stored in the RAM.

This work aims to evolve an adaptive controller capable of tolerating certain robot faults using CHE. The main feature is the use of GA pipeline which plays a crucial role in the hardware evolution. In this approach, the control hardware allows the communication with external computers or prototype boards. Furthermore, external signals are connected to the evolution designs to conduct fitness evaluation of the current generation, and the results are fed into the GA pipeline for further manipulations.

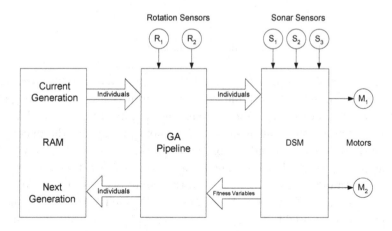

Fig. 2.4 The complete hardware evolution.

C. Keymeulen et al.: Model-free off-line and model-based on-line hard-ware evolution

Keymeulen et al. (1998a, 1998b, 1998c, 1999) evaluated two approaches of implementing EHW onto a robot navigation system, i.e., off-line model-free and on-line model-based evolutions. The task of the robot is to track a red ball while avoiding obstacles during its motion. As shown in Fig. 2.5, the model-free method simulates the evolution process of the robot controller in an artificial environment. GA was used to search off-line using the training data to achieve robust robot behavior for maintaining the robot's performance in the real-world. In this approach, both the EHW and the environment were simulated to find a controller that is capable of accomplishing the specified task. The best controller found by the evolution was then used to control the real robot in a physical environment. In this context, the GA is used as an optimization strategy to find the optimal controller for a given simulated environment and consequently the population size adopted is relatively small.

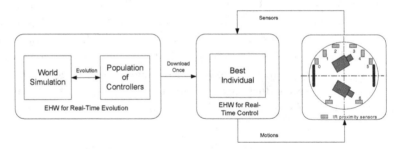

Fig. 2.5   The model-free evolution for a real robot.

The model-free approach assumed that the simulated worlds are carefully designed and the evolvable Boolean controllers are robust enough to deal with the uncertainties and inaccuracies of motor responses and sensor values. However, the simulated world may not approximate the real-world very well in most cases and the on-line evolution is often extremely time-consuming. A model-based on-line evolution scheme was thus proposed to reduce the number of interactions with the environment and yet maintains a good on-line performance. In on-line model-based hardware evolution, the environment used to train the robot controller is modeled in real-time. As shown in Fig. 2.6, two EHWs are used to execute the robot behaviors and to simulate the evolution process, respectively. At each time step, the on-line model-based evolution simulates the evolution process of robot controllers in the current world model of the environment to compute the optimal controller. After a few generations, the best controller found by the evolution is downloaded to the EHW to control the robot in the real-world. In this approach, the learning and execution phases are concurrent, and the GA is used as an adaptive strategy to find the optimal controller continuously for an ever-changing world model.

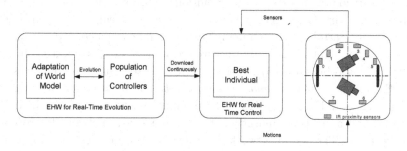

Fig. 2.6   The model-based evolution for a real robot.

D. Lipson and Pollack: Automatic design and manufacture of robotic lifeforms

At the macro hardware level, Lipson and Pollack (2000) proposed a novel approach to evolve real robots by combining evolution simulations with rapid manufacturing technology. In this way the electromechanical objects are evolved by evolutionary computation and fabricated robotically using rapid prototyping technology. Fig. 2.7 illustrates the associated evolution mechanism for deriving the structure and control of a robot.

In this scheme, bars and actuators are used as building blocks of structure, and artificial neurons as building blocks of control. Bars connected with free joints can potentially form trusses that represent arbitrary rigid, flexible, and articulated structures and emulate revolute, linear, and planar joints at various levels of the hierarchy. Similarly, sigmoidal neurons can connect to create arbitrary control architectures, such as feed-forward and recurrent nets, state machines, and multiple independent controllers. During the evolution simulations, some substructures may become rigid, while others may become articulated. A sequence of operators will construct a robot and its controller from scratch by adding, modifying and removing building blocks. In the sequence at the bottom of Fig. 2.7, the derivation of a bar and its neuron control are illustrated.

A simulated evolution process was performed using GOLEM (Genetically Organized Lifelike Electro-Mechanics), where the fitness function is defined as the net Euclidean distance that the center-of-mass of an individual moves over a fixed number of cycles of its neuron control. The population is comprised of 200 null empty individuals and is manipulated by genetic operations during the evolution. The evolution process was continued for 300 to 600 generations to derive the desired electromechanical objects. Prior to this work, the evolutionary robotics involved either entirely in the virtual world or, when it was applied in reality, the adaptation of only the control level of manually designed and constructed robots which have predominately fixed architecture. However, in this novel evolution scheme, real robots are evolved from both their structures and controls.

E. Floreano and Mondada: Evolvable robots based on neural control

Floreano and Mondada (1994; 1996) employed a standard GA with fitness scaling and biased mutations to evolve a neural network for robotic applications. The neural network synaptic weights and thresholds are coded on chromosomes in a population, and each chromosome is in turn decoded into a neural network. The neural network is connected to eight infrared proximity sensors and two motors. The robot was let free to move for a certain amount of time while its fitness value was recorded and stored. After all individuals in the population have been evaluated, the three genetic operations including selection, crossover and mutation were applied to reproduce a new population of the same size. In the experiment, the Khepera robot was put in a maze with irregularly shaped walls. The fitness function was designed to select good individuals that could keep a low sensor activity and high motor activation. The system was evolved for 100 generations before behaviors of smooth navigation and efficient obstacle avoidance were obtained. Fig. 2.8 shows the neural structure (left) and the structure of the neural networks under evolution (right). From the evolution results, Floreano and Mondada (1994; 1996) observed that all individuals in the population had developed a preferential direction of motion corresponding to the side where more sensors were located.

In this research, the evolution is carried out on a real robot and the goal is to exploit the interaction with real environment during the evolution process. In autonomous robots, it is rather difficult to employ strictly supervised learning algorithms because the exact system output is not always available. The results in this experiment show that the evolutionary approach is well suited for deriving the desired robot behaviors in a real environment. The continuous interaction with the environment and the simultaneous control structure evolution generate the desired robotic system with satisfactory behaviors. However, this approach is often time-consuming and thus it is hard to scale up to other complex and practical robotic applications.

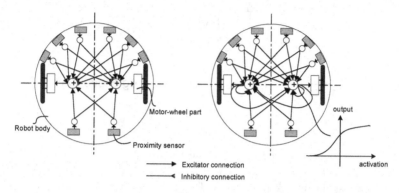

Fig. 2.7   The neural network structure for Khepera robot evolution.

F. Lewis, Fagg and Bekey: Evolution of gait patterns in walking robots
The goal of the work by Lewis, et al., (1994) is to evolve a neural network to generate a sequence of signals driving the legs of a hexapod robot to produce consistent forward locomotion along the body axis. The position of joint is driven by the state of a neuron. The two neurons that are associated with a particular leg are implemented as an oscillator circuit. Fig. 2.9 shows the local leg circuit. By setting the appropriate weight (W01 and W10) and threshold (T0 and T1) parameters, the two neurons will oscillate at a particular frequency and phase.

The work applied the staged evolution to derive the desired walking gait for a six-legged robot. The first phase of the experiment was to accomplish the oscillator circuit evolution. The genetic code specifies the four parameters involved in this circuit, i.e., the two weights and the two thresholds. Performance evaluation was conducted by observing the temporal behavior of the oscillator neurons. When the neural network reaches the state of consistent oscillations, the experiment will be transited to the next phase, where the task is to evolve the desired walking gait by setting the connections between the oscillators. As shown in Fig. 2.10, four additional parameters are needed to specify these connections.

The genetic chromosome used in this experiment consists of 65 bits, where 8 bits are used for each of the 8 parameters, i.e., W01, W10, T0,

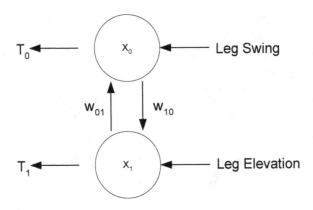

Fig. 2.8   An oscillator circuit.

T1 (single oscillator circuit) and A, B, C, D (oscillator interconnections). Each parameter encodes a weight value in the range of [-8, 8] using a gray coding approach. The final bit (65th) determines the mapping from the leg swing neuron to the joint actuator. At the end of the evolution, the robot was learned to produce a tripod gait where the left-forward and backward legs moved in phase with the right-middle leg, i.e., at least three legs are on the ground at any time. This gait is common in insects. The evolution results show that the convergence rate had been improved by employing GA to adjust the interconnection weights in the neural networks. This staged evolution approach can also be applied to other robotic control applications.

2.5   Promises and Challenges

2.5.1   *Promises*

As a promising research field, evolvable hardware is capable of bringing novel approaches to robot controller designs that cannot be produced efficiently by conventional approaches often adopted by human designers [Layzell, 1998]:

- The intrinsic evolution process actively exploits side effects and obscures physical phenomena of the evolvable hardware to improve the fitness value for more and better robotic behaviors. Furthermore, the evolution is free to explore very unusual circuits beyond the conventional designs, which may provide better solutions for the robotic control system.

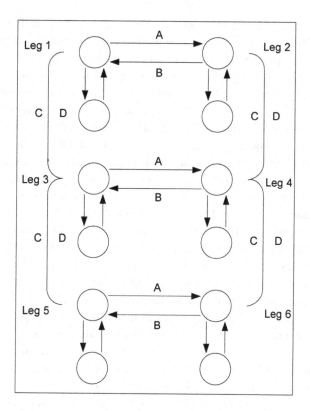

Fig. 2.9 The coupled oscillator circuits.

- Complex behaviors such as robot navigation in dynamic environments can be evolved with only a few hardware components. For example, Thompson (1995a, 1996a) used the configuration of a small corner of XC6212 FPGA to derive a simple anti-collision robotic controller. The indispensable components in conventional designs such as system clock can also be under the evolutionary control. Therefore the evolved system can be more compact with less power consumption than the conventional ones.
- EAs can be used in the field of fault tolerance that is important for autonomous robot navigation. In such system, the robot is expected to run in a noisy environment where robot sensor and motor faults may occur. Therefore, adaptive component failure compensation is desired. For example, in space exploration the radiation and high temperature tolerance

based devices are urgently needed. Since EAs are generally robust and not sensitive to noise, they can be used to evolve robotic systems capable of tolerating the occurred faults and system uncertainties.

- The robot controller designs based on evolvable hardware can be achieved with little knowledge of electronics. It is possible for non-electronics engineers to design circuits for robotic control systems using EHW, since in this approach the operations are often based on the robotic behaviors instead of the control circuits. The evolved circuit can be thought of as a black-box and only its inputs and outputs are meaningful for the designers. Therefore in the design process more attentions can be focused to the design strategies instead of the circuit configurations.

- Conventional GA and GP are not suitable for long-term open-ended incremental evolution of complex systems such as autonomous robot navigation. Harvey's Species Adaptation Genetic Algorithm (SAGA) (1992) was developed to deal with this issue. In SAGA, different structures are coded in genotypes of different lengths, which give a search space of open-ended dimensionality. Therefore SAGA should be a suitable choice for incremental evolution applications like autonomous robot navigation.

- At the macro hardware level, technology advances in micro-electro-mechanic system (MEMS), nanofabrication, and multi-material rapid prototyping, etc. may pave the way for evolving more complex and self-organizing structures (Lipson and Pollack, 2000). The increasing interests in these areas have been observed in recent evolvable hardware conferences.

2.5.2   *Challenges*

There are certain problems in EHW-based robotic control systems that remain open so far. Several important issues such as portability, scalability, and building blocks of EHW should be fully investigated before the EHW could be widely applied to real-world robotic applications [Layzell, 1998]:

- Both the intrinsic and extrinsic EHWs suffer from similar problems in the portability of the evolved circuits. Many existing works were produced in simulations, which are often difficult to be physically implemented in real hardware. The reliable transfer from simulation to physical hardware implementation without any loss of robotic behaviors is an important issue for future research in this field.

- An important research area for exploration is the complete robotic behaviors by means of on-line intrinsic EHW that is difficult to be achieved particularly for real-world robotic applications. This approach is alluring since on-line evolution is often hard to be realized by conventional design methodologies, and could be an advantage of EHW over other artificial intelligence approaches in robotic applications.

- Although a number of tactics have been used to analyze the exotic evolved circuits (Thompson and Layzell, 1999), unconventional circuits are still hard to analyze as compared to the more orthodox ones. More effective and novel techniques should be developed to tackle the problem of circuitry analysis, so that human beings can derive the desired robotic behaviors more deliberately by using the innovative circuit design methods inspired by the evolved circuitry analysis.

- The reliability of the evolved circuits should also be considered. Due to the dependency of circuit components on external environments like temperature, the evolved circuit may not work well when external factors vary (Thompson, 1996d; 1997b). Moreover, the evolution strategy that is successfully implemented in an EHW chip may cease to work in another similar type of EHW chip, even in the same batch. Therefore experimental results are sometimes hard to be repeated by other researchers using the same facilities.

- In many existing works, only small circuits have been evolved successfully. Complex robotic behaviors require the evolution of large controller circuits, e.g., the binary representation often adopted in GA could lead to undesired huge chromosome length in complex applications. Although function-level EHW allows complex circuits to be defined, it could result in less continuous dynamics in the system. Undoubtedly, more research efforts are needed in this direction before EHW could make a significant impact on real-world robotic applications.

2.6 Summary

The application of evolvable hardware in evolutionary robotics has been discussed, and a few representative experiments in the field have been reviewed in this chapter. In addition, the context of evolvable hardware and evolutionary robotics has been reviewed respectively. The potentialities and limitations of the EHW-based robotic systems have also been discussed and summarized in this chapter. To develop EHW for real-world robotic applications to a level where they can be seriously considered as a viable alternative to the more orthodox approaches, the challenges discussed in this chapter need to be addressed. Some killer robotic applications via EHW which cannot be effectively implemented by other approaches should also be accomplished. It is believed that more attention and interests from researchers in various disciplines will be expected for this promising research field in the years to come.

Bibliography

Arsene, C. T. C., and Zalzala, A. M. S. (1999). Control of autonomous robots using fuzzy logic controllers tuned by genetic algorithms, *Proceedings of the IEEE Congress on Evolutionary Computation*, pp. 428–435.

Barfoot, T. D., and D'Eleuterio, G. M. T. (1999). An evolutionary approach to multiagent heap formation, *Proceedings of the IEEE Congress on Evolutionary Computation*, pp. 420–427.

Bennett III, F. H., Koza, J. R., Andre, D., Keane, M. A. (1996). Evolution of a 60 decibel op amp using genetic programming, Evolvable Systems: From Biology to Hardware, *Lecture Notes in Computer Science 1259 (ICES96)*, Springer-Verlag, pp. 455–469.

Berlanga, A., Sanchis, A., Isasi, P., Molina, J. M. (1999). Neural networks robot controller trained with evolution strategies, *Proceedings of the IEEE Congress on Evolutionary Computation*, pp. 413–419.

Brooks, R. (1992). Artificial life and real robots, In F. J. Varela and P. Bourgine (Eds.), *Proceedings of the 1st European Conference on Artificial Life*, Cambridge MA: MIT Press/Bradford Books, pp. 3–10.

Brooks, R. (2000). From robot dreams to reality, *Nature*, vol. 406, pp. 945–947.

Chaiyaratana, N. and Zalzala, A. M. S. (1999). Hybridisation of neural networks and genetic algorithms for time-optimal control, *Proceedings of the IEEE Congress on Evolutionary Computation*, pp. 389–396.

Chocron, O. and Bidaud, P. (1999). Evolving walking robots for global task based design, *Proceedings of IEEE Congress on Evolutionary Computation*, pp. 405–412.

de Garis, H. (1996). CAM-BRAIN: The evolutionary engineering of a billion neuron artificial brain by 2001 which grows/evolves at electronic speeds inside a Cellular Automation Machine (CAM), *In Sanchez, E., and Tomassini, M. (Eds.), Towards Elolvable Hardware: The evolutionary engineering approach, vol. 1062 of LNCS*, Springer-Verlag, pp. 76–98.

Dittrich, P., Burgel, A. and Banzhaf., W. (1998). Learning to move a robot with random morphology, *Proceedings of the 1st European Workshop on Evolutionary Robotics 98 (EvoRobot98)*, France, pp. 165–178.

Dorigo, M., and Colombetti, M. (1994). Robot shaping: Developing autonomous agent through learning, *Artificial Intelligence*, vol. 71, pp. 321–370.

Dorigo, M., and Schnepf, U. (1993). Genetics-based machine learning and behavior-based robotics: A new synthesis, *IEEE Transactions on Systems, Man, Cybernetics*, vol. 23, pp. 141–154.

Earon, E. J. P., Barfoot, T. D., and D'Eleuterio, G. M. T. (2000). From the sea to the sidewalk: the evolution of hexapod walking gaits by a genetic algorithm, *Evolvable Systems: From Biology to Hardware, Lecture Notes in Computer Science 1801 (ICES2000)*, Springer-Verlag, pp. 51–60.

Ficici, S. G., and Pollack, J. B. (1999). Embodied evolution: embodying an evolutionary algorithm in a population of robots, *Proceedings of the IEEE Congress on Evolutionary Computation*, pp. 335-342.

Floreano, D., and Mondada, F. (1994). Automatic creation of an autonomous agent: Genetic evolution of a neural-network driven robot, *Proceedings of the 3rd International Conference on Simulation of Adaptive Behavior*, Cambridge, MA: MIT Press, pp. 421–430.

Floreano, D., and Mondada, F. (1996). Evolution of homing navigation in a real mobile robot, *IEEE Transactions on Systems, Man, and Cybernetics: Part B (Cybernetics)*, vol. 26, no. 3, pp 396–407.

Floreano, D., and Urzelai, J. (2000), Evolutionary robots with on-line self-organization and behavioral fitness,*Neural Networks*, vol. 13, pp. 431–443.

Fogel, D. B. (2000). What is evolutionary computation? *IEEE Spectrum*, pp. 26–32.

Forbes, N. (2000). Life as it could be: Alife attempts to simulate evolution, *IEEE Intelligent Systems*, December 2000, pp. 1–6.

Goeke, M., Sipper, M., Mange, D., Stauffer, A., Sanchez, E., Tomassini, M. (1997). On-line autonomous evolware, *Evolvable Systems: From Biology to Hardware, Lecture Notes in Computer Science 1259 (ICES96)*, Springer-Verlag, pp. 96–106.

Goldberg, D. E. (1989). *Genetic Algorithms in Search, Optimization and Machine learning*, Addison Wesley.

Gomi, T. and Griffith, A. (1996). Evolutionary robotics - an overview, *Proceedings of the 3rd International Conference on Evolutionary Computation*, Nagoya, Japan, pp. 40–49.

Grefenstette, J. J., and Schultz, A. (1994). An evolutionary approach to learning in robots, *Proceedings of the Machine Learning Workshop on Robot Learning, 11th International Conference on Machine Learning*, Morgan Kaufmann, New Brunswick, N.J., pp. 65–72.

Grefenstette, J. J. (1996). Genetic learning for adaptation in autonomous robots,*Robotics and manufacturing: Recent Trends in Research and Applications*. vol. 6, ASME Press, New York, pp. 265–270.

Gruau, F., and Quatramaran, K. (1997). Cellular encoding for interactive evolutionary robotics, *Proceedings of the Fourth European Conference on Artificial Life*, The MIT Press, pp. 368–377.

Haddow P., and Tufte, G. (1999). Evolving a robot controller in hardware, *Proceedings of Norwegian Computer Science Conference (NIK99)*, pp. 141–150.

Haddow, P. and Tufte, G. (2000). An evolvable hardware FPGA for adaptive hardware, *Proceedings of the IEEE Congress on Evolutionary Computation*, pp. 553–560.

Hamilton, A., Papathanasiou, K., Tamplin, M. R., and Brandtner, T. (1998). Palmo: Field programmable analog and mixed-signal VLSI for evolution hardware, *Evolvable Systems: From Biology to Hardware, Lecture Notes in Computer Science 1478 (ICES1998)*, Springer-Verlag, pp. 335–344.

Harvey, I. (1992). Species adaptation genetic algorithms: the basis for a continuing SAGA, *Proceedings of the 1st European Conference on Artificial Life*, MIT Press/Bradford Books, pp. 346–354.

Hasegawa, Y., Mase, K. and Fukuda, T. (1999). Re-grasping behavior acquisition by evolution programming, *Proceedings of the IEEE Congress on Evolutionary Computation*, pp. 374–380.

Hemmi, H., Mizoguchi, J., and Shimohara, K. (1994). Development and evolution of hardware behaviors, In Brooks, R., and Maes, P., (Eds.), *Artificial Life IV: Proceeding of the 4th International Workshop on the Synthesis and Simulation of Living Systems*, MIT Press, pp. 371–376.

Higuchi, T., Niwa, T., Tanaka, T., Iba, H., Garis, H. d., and Furuya, T. (1993). Evolving hardware with genetic learning: a first step towards building a Darwin machine, *Proceeding of the 2nd International Conference on From Animals to Animats 2: Simulation of Adaptive Behavior*, Honolulu, Hawaii, United States, pp. 417–424.

Higuchi, T., Iwata, M., Kajitani, I., Murakawa, M., Yoshizawa, S. and Furuya T. (1996a). Hardware evolution at gate and function levels, *Proceeding of Biologically Inspired Autonomous Systems: Computation, Cognition and Action*.

Higuchi, T., Iwata, M., Kajitani, I., Yamada, H., Manderick, B., Hirao, Y., Murakawa, M., Yoshizawa, S., Furuya, T. (1996b). Evolvable hardware with genetic learning, *Proceeding of IEEE International Symposium on Circuits and Systems (ISCAS96)*, vol. 4, pp. 29–32.

Higuchi, T. Iwata, M. Kajitani, I. Iba, H., Hirao, Y., Manderick, B., and Furuya, T. (1997a). Evolvable hardware and its applications to pattern recognition and fault-tolerant systems, *Towards Evolvable Hardware: the evolutionary engineering approach, Lecture Notes in Computer Science 1062*, Springer-Verlag. pp. 118–135.

Higuchi, T. Murakawa, M., Iwata, M., Kajitani, I., Liu, W., and Salami, M. (1997b). Evolvable hardware at function level, *Proceeding of IEEE International Conference on Evolutionary Computation*, pp. 187–192.

Higuchi, T., and Kajihara, N. (1999). Evolvable hardware chips for industrial applications, *Communications of the ACM*, vol. 42, no. 4, pp. 60–69.

Higuchi, T., Iwata, M., Keymeulen, D., Sakanashi, H., Murakawa, M., Kajitani, I., Takahashi, E., Toda, K., Salami, M., Kajihara, N., and Otsu, N. (1999). Real-world applications of analog and digital evolvable hardware, *IEEE Transactions on Evolutionary Computation*, vol. 3, no. 3, pp. 220–235.

Hirst, A. J. (1996). Notes on the evolution of adaptive hardware, In Parmee, I. (Ed.), *Proceeding of second International Conference on Adaptive Computing in Engineering Design and Control*, University of Plymouth, UK, pp.

212–219.

Holland, J. H. (1975). *Adaptation In Natural And Artificial Systems*, Ann Arbor: The University of Michigan Press.

Huang, S. H. (2002). Artificial Neural Networks and Its Manufacturing Application: Part I,
http://www.min.uc.edu/icams/resources/ANN/ANNManuI.ppt.

Husbands, P., Harvey, I., Cliff, D., Miller, G. (1994). The use of genetic algorithms for the development of sensorimotor control systems, In Nicoud and Gaussier (Eds.). *From Perception to Action*, IEEE Computer Society Press, pp. 110–121.

Iba, H., Iwata, M., and Higuchi, T. (1997). Machine learning approach to gate-level evolvable hardware, *Evolvable Systems: From Biology to Hardware, Lecture Notes in Computer Science 1259 (ICES96)*, Springer-Verlag, pp.327–343.

Iwata, M., Kajitani, I., Yamada, H., Iba, H., and Higuchi, T. (1996). A pattern recognition system using evolvable hardware, *Parallel Problem Solving from Nature - PPSN IV, Lecture Notes in Computer Science 1141*, Springer-Verlag, pp.761–770.

Iwata, M., Kajitani, I., Murakawa, M., Hirao, Y., Iba, H., Higuchi, T. (2000). Pattern recognition system using evolvable hardware, *Systems and Computers in Japan*, vol. 31, no. 4, pp. 1–11.

Jakobi, N. (1997). Half-baked, ad-hoc, and noisy: minimal simulations for evolutionary robotics, In Husbands, P., and Harvey, I., (Eds.) *Proceeding of the Forth European Conference on Artificial Life*, MIT Press, pp. 348–357.

Jakobi, N., and Quinn, M. (1998). Some problems (and a few solutions) for open-ended evolutionary robotics, *Proceedings of the First European Workshop on Evolutionary Robotics 98 (EvoRobot98)*, France, pp. 108-122.

Keramas, J. G. (2000). How will a robot change your life? *IEEE Robotics and Automation Magazine*, March, pp. 57–62.

Keymeulen, D. Konaka, K., Iwata, M., Kuniyoshi, Y., and Higuchi, T. (1997). Robot learning using gate-level evolvable hardware, *Proceedings of the Sixth European Workshop on Learning Robots (EWLR)*, Springer-Verlag.

Keymeulen, D., Iwata, M., Konaka, D., Suzuki, R., Kuniyoshi, Y., and Higuchi, T. (1998a). Off-line model-free and on-line model-based evolution for tracking navigation using evolvable hardware, *Proceedings of the 1st European Workshop on Evolutionary Robotics*, Springer-Verlag, pp. 208–223.

Keymeulen, D., Iwata, M., Kuniyoshi, Y., Higuchi, T. (1998b). On-line model-based learning using evolvable hardware for a robotics tracking systems, Koza et al. (Eds.), *Genetic Programming 1998: Proceeding of the 3rd Annual Conference*, Morgan Kaufmann, pp. 816–823.

Keymeulen, D., Iwata, M., Kuniyoshi, Y., Higuchi, T. (1998c). Comparison between an off-line model-free and an on-line model-based evolution applied to a robotics navigation system using evolvable hardware, *Artificial Life VI: Proceedings of the 6th International Conference on Artificial Life*, MIT Press, pp. 109–209.

Keymeulen, D., Iwata, M., Kuniyoshi, Y., Higuchi, T. (1999). On-line evolution for a self-adapting robotic navigation system using evolvable hardware, *Artificial Life Journal*, vol. 4, no. 4, pp. 359–393.

Kitano, H. (1996). Challenges of evolvable systems: analysis and future directions, *Evolvable Systems: From Biology to Hardware, Lecture Notes in Computer Science 1259 (ICES96)*, Springer-Verlag, pp. 125–135.

Kondo, T., Ishiguro, A., Tokura, S., Uchikawa, Y., and Eggenberger, P. (1999). Realization of robust controllers in evolutionary robotics: a dynamically-rearranging neural network approach, *Proceedings of the IEEE Congress on Evolutionary Congress*, pp. 366–373.

Koza, J. R. (1994). *Genetic Programming II*, The MIT Press, Cambridge, Mass., USA.

Koza, J. R., Andre, D., Bennett III, F. H., and Keane, M. A. (1996). Use of automatically defined functions and architecture-altering operations in automated circuit synthesis using genetic programming, *Proceeding of the Genetic Programming 1996 Conference (GP96)*, The MIT Press, pp. 132–149.

Kubota, N., Morioka, T., Kojimi, F., and Fukuda, T. (1999). Perception-based genetic algorithm for a mobile robot with fuzzy controllers, *Proceedings of the IEEE Congress on Evolutionary Computation*, pp. 397–404.

Layzell, P. (1998). The evolvable motherboard: a test platform for the research of intrinsic hardware evolution, *Cognitive Science Research Paper 479*.

Layzell, P. (1999). Reducing hardware evolution's dependency on FPGAs, *Proceedings of the 7th International Conference on Microelectronics for Neural, Fuzzy, and Bio-Inspired Systems (MicroNeuro99)*, pp. 171–178.

Lazarus, C. and Hu, H. (2001). Using genetic programming to evolve robot behaviors, *Proceedings of the 3rd British Conference on Autonomous Mobile Robotics and Autonomous Systems*, Manchester.

Lewis, M. A., Fagg, A. H., and Solidum, A. (1992). Genetic programming approach to the construction of a neural network for control of a walking robot, *Proceedings of IEEE International Conference on Robotics and Automation*, France, pp. 2618–2623.

Lewis, M. A., Fagg, A. H., and Bekey, G. A. (1994). Genetic algorithms for gait synthesis in a hexapod robot, *Recent Trends in Mobile Robots*, (Zhang, Ed.), World Scientific, New Jersey, pp. 317–331.

Lipson, H. and Pollack, B. (2000). Automatic design and manufacture of robotic platforms, *Nature*, vol. 46, August 2000, pp. 974–977.

Liu, J., Pok, C. K., and Keung, H. K. (1999). Learning coordinated maneuvers in complex environments: a sumo experiment, *Proceedings of the IEEE Congress on Evolutionary Computation*, pp. 343–349.

Lohn, J. D. and Colombano, S. P. (1998). Automated analog circuit synthesis using a linear representation, *Evolvable Systems: From Biology to Hardware, Lecture Notes in Computer Science 1478 (Proc. of ICES1998)*, Springer-Verlag, pp. 125–133.

Lund, H. H., and Miglino, O. (1996). From simulated to real robots, *Proceedings of the IEEE Congress on Evolutionary Computation*, pp. 362–365.

Manderick, B. and Higuchi, T. (1996). Evolvable hardware: an outlook, *Evolvable Systems: From Biology to Hardware, Lecture Notes in Computer Science*

1259 (ICES96), Springer-Verlag, pp. 305–310.

Meyer, J.-A., Husbands, P., and Harvey, I. (1998). Evolutionary robotics: A survey of applications and problems, *Proceedings of the 1st European Workshop on Evolutionary Robotics 98 (EvoRobot98)*, France, pp. 1–21.

Michalewicz, Z. (1994). *Genetic Algorithms + Data Structures = Evolutionary Programsm*, Springer-Verlag.

Miglino, O., Denaro, D., Tascini, G., and Parisi, D. (1998). Detour behavior in evolving robots: Are internal representations necessary? *Proceedings of the First European Workshop on Evolutionary Robotics 98 (EvoRobot98)*, France, pp. 59–70.

Mizohuchi, J., Hemmi, H., and Shimohara, K. (1994). Production genetic algorithms for automated hardware design through an evolutionary process, In Michalewicz, Z. (Ed.), *Proceeding of the First IEEE Conference on Evolutionary Computation*, pp. 661–664.

Mondada, F. and Floreano, D. (1995) Evolution of neural control structures: Some experiments on mobile robots, *Robotics and Autonomous Systems*, vol. 16, pp. 183–195.

Naito, T., Odagiri, R., Matsunaga, Y., Tanifuji, M. and Murase, K. (1997). Genetic evolution of a logic circuit which controls an autonomous mobile robot, *Evolvable Systems: From Biology to Hardware, Lecture Notes in Computer Science 1259 (ICES96)*, Springer-Verlag, pp. 210–219.

Nolfi, S., Floreano, D., Miglino, O., and Mondada, F. (1994). How to evolve autonomous robots: Different approaches in evolutionary robotics, In R. Brooks and P. Maes, (eds.), *Artificial Life IV*, MIT Press/Bradford Books, pp. 190–197.

Nolfi, S., and Floreano, D. (2000). *Evolutionary robotics: biology, intelligence, and technology of self-organizing machines*, Cambridge, MA: MIT Press.

Nordin, P. and Banzhaf, W. (1997). Real-time control of a Khepera robot using genetic programming, *Control and Cybernetics*, vol. 26, no. 3, pp. 533–561.

Nordin, P., Banzhaf, W., and Brameier, M. (1998). Evolution of a world model for a miniature robot using genetic programming, *Robotics and Autonomous Systems*, vol. 25, pp. 105–116.

Odagiri, R., Yu, W., Asai, T., and Murase, K. (1998). Analysis of the scenery perceived by a real mobile robot Khepera, *Evolvable Systems: From Biology to Hardware, Lecture Notes in Computer Science 1478 (ICES1998)*, Springer-Verlag, pp. 295–302.

Pollack, J. B., Lipson, H., Ficci, S., Funes, P., Hornby, G., and Watson, R. (2000). Evolutionary techniques in physical robotics, *Evolvable Systems: From Biology to Hardware, Lecture Notes in Computer Science 1801*, Springer-Verlag, pp. 175–186.

Pratihar, D. K., Deb, K., and Ghosh, A. (1999). Fuzzy-genetic algorithms and mobile robot navigation among static obstacles, *Proceedings of the IEEE Congress on Evolutionary Computation*, pp. 327–334.

Quarles, T., Newton, A. R., Pederson, D. O., and Sangiovanni-Vincentelli, A. (1994). *SPICE 3 Version 3F5 User's Manual*, Department of Electrical Engineering and Computer Science, University of California, Berkeley, CA,

USA.

Revello, T. E., and McCartney, R. (2000). A cost term in an evolutionary robotics fitness function, *Proceedings of the IEEE Congress on Evolutionary Computation*, pp. 125–132.

Reynolds, C. W. (1994). An evolved, vision-based model of obstacle avoidance behavior, *Artificial Life III*, Reading, MA: Addison-Wesley, pp. 327–346.

Ros, H. (1997). Evolutionary strategies of optimization, *Phys. Rev. E*, vol. 56, pp. 1171–1180.

Sakanashi, H., Higuchi, T., Iba, H., and Kakazu, Y. (1996). Evolution of binary decision diagrams for digital circuit design using genetic programming, *Evolvable Systems: From Biology to Hardware, Lecture Notes in Computer Science 1259 (ICES96)*, Springer-Verlag, pp. 470–481.

Sakanashi, H., Tanaka, M., Iwata, M., Keymulen, D., Murakawa, M., Kajitani, T., Higuchi, T. (1999). Evolvable hardware chips and their applications, *Proceeding of the IEEE Systems, Man, and Cybernetics Conference (SMC99)*, pp. V559–V564.

Sanchez, E., Mange, D., Sipper, M., Tomassini, M., Prez-Uribe, A., and Stauffer, A. (1996), Phylogeny, ontogeny, and epigenesis: three sources of biological inspiration for softening hardware, *Evolvable Systems: From Biology to Hardware, Lecture Notes in Computer Science 1259 (ICES96)*, Springer-Verlag, pp. 35–54.

Schaefer, C. G. (1999). Morphogenesis of path plan sequences through genetic synthesis of L-system productions, *Proceedings of the IEEE Congress on Evolutionary Computation*, pp. 358–365.

Shirasuchi, S. (1996). FPGA as a key component for reconfigurable system, *Evolvable Systems: From Biology to Hardware, Lecture Notes in Computer Science 1259 (ICES96)*, Springer-Verlag, pp. 23–32.

Sipper, M. (1996). Designing evolware by cellular programming, *Evolvable Systems: From Biology to Hardware, Lecture Notes in Computer Science 1259 (ICES96)*, Springer-Verlag, pp. 81–95.

Sipper, M. and Ronald, E. M. A. (2000). A new species of hardware, *IEEE Spectrum*, March 2000, pp. 59–64.

Stauffer, A. and Sipper, M. (1998). Modeling cellular development using L-systems, *Evolvable systems: From Biology to Hardware, Lecture Notes in Computer Science 1478*, Springer-Verlag, pp. 196–205.

Stoica, A., Zebulum, R., and Keymeulen, D. (2000). Mixtrinsic evolution, *Evolvable systems: from biology to hardware, Lecture Notes in Computer Science 1801*, Springer-Verlag, pp. 208–217.

Tan, K. C., Lee, T. H., Khoo, D. and Khor, E. F. (2001). A multi-objective evolutionary algorithm toolbox for computer-aided multi-objective optimization, *IEEE Transactions on Systems, Man and Cybernetics: Part B (Cybernetics)*, vol. 31, no. 4, pp. 537–556.

Thompson, A. (1995a). Evolving electronic robot controllers that exploit hardware resources, *Proceeding of the 3rd European Conference on Artificial Life*, Springer-Verlag, pp. 640–656.

Thompson, A. (1995b). Evolving fault tolerance systems, *Proceeding of the 1st*

IEE/IEEE International Conference on Genetic Algorithms in Engineering Systems: Innovations and Applications, pp. 524–529.

Thompson, A. (1996a). An evolved circuit, intrinsic in silicon, entwined with physics, *Evolvable Systems: From Biology to Hardware, Lecture Notes in Computer Science 1259 (ICES96)*, Springer-Verlag, pp. 390–405.

Thompson, A. (1996b). Silicon evolution, In Koza, J. R., et al. (eds.), *Genetic Programming 1996: Proceeding 1st Annual Conference (GP96)*, Cambridge, MA: MIT Press, pp. 444–452.

Thompson, A. (1996c). Evolutionary techniques for fault tolerance, *Proceedings of UKACC International Conference on Control*, pp. 693–698.

Thompson, A. (1996d). *Hardware evolution: Automatic design of electronic circuits in reconfigurable hardware by artificial evolution*, Ph.D. thesis, University of Sussex, UK.

Thompson, A. (1997a). Evolving inherently fault-tolerant systems, *Proceeding of the Institute of Mechanical Engineers, Part I: Journal of Systems and Control Engineering*, vol. 211, pp. 365–371.

Thompson, A. (1997b). Temperature in Natural and Artificial Systems, *Proceeding 4th European Conference on Artificial Life*, Husbands, P. and Harvey, I. (eds.), MIT Press, pp. 388–397.

Thompson, A. and Layzell, P. (1999). Analysis of Unconventional Evolved Electronics, *Communications of the ACM: Special Section of Communications of the ACM*, Xin Yao (Ed.), vol. 42, no. 4, pp. 71–79.

Thompson, A., Layzell, P., and Zebulum, R. S. (1999). Explorations in design space: unconventional electronics design through artificial evolution, *IEEE Transactions on Evolutionary Computation*, vol. 3, no. 3, pp. 167–196.

Walker, J. F., and Oliver, J. H. (1997). A survey of artificial life and evolutionary robotics, http://citeseer.nj.nec.com/walker97survey.html. Watson, R. A.,

Xilinx Inc. (1996). *XC6200 Product Specification V1.0*, 1996, http://www.xilinx.com.

Yamada, S. (1998). Learning behaviors for environmental modeling by genetic algorithm, *Proceedings of the 1st European Workshop on Evolutionary Robotics 98 (EvoRobot98)*, France, pp. 179–191.

Yamamoto, J., and Anzai, Y. (1997). Autonomous robot with evolving algorithm based on biological systems, *Evolvable Systems: From Biology to Hardware, Lecture Notes in Computer Science 1259 (ICES96)*, Springer-Verlag, pp. 220–233.

Yao, X. (1999). Following the path of evolvable hardware, *Communications of the ACM*, vol. 42, no. 4, pp. 47–49.

Yao, X. and Higuchi, T. (1999). Promises and challenges of evolvable hardware, *IEEE Trans. on Systems, Man, and Cybernetics- Part C: Applications and Reviews*, vol. 29, no. 1, pp. 87–97.

Zebulum, R. S., Pacheco, M. R., and Vellasco, M. (1996). Evolvable systems in hardware design: taxonomy, survey and applications, *Evolvable Systems: From Biology to Hardware, Lecture Notes in Computer Science 1259 (ICES96)*, Springer-Verlag, pp. 344–358.

Chapter 3

FPGA-Based Autonomous Robot Navigation via Intrinsic Evolution*

In the recent decades researchers have been working on the application of artificial evolution in mobile robots for adapting their behaviors in unstructured environments continuously and autonomously. At the same time great interests have been shown in the development of evolvable hardware which is a new set of integrated circuits capable of reconfiguring their architectures using evolutionary computing techniques. This chapter presents the design and real-time implementation of an evolvable hardware based autonomous robot navigation system using intrinsic evolution. Distinguished from the traditional evolutionary approaches based on software simulation, an evolvable robot controller at the hardware gate-level that is capable of adapting dynamic changes in the environments is implemented. In our approach, the concept of Boolean function is used to construct the evolvable controller implemented on an FPGA-based robot turret, and evolutionary computing is applied as a learning tool to guide the artificial evolution at the hardware level. The effectiveness of the proposed evolvable autonomous robotic system is confirmed with the physical real-time implementation of robot navigation behaviors on light source following and obstacle avoidance using a robot with traction fault.

3.1 Introduction

In the recent decades researchers have been working on the application of artificial evolution to autonomous mobile robots for adapting their behaviors to changes in an environment continuously. As anticipated by Keramas

*Copyright (2004) From (Portions of "FPGA-based automomous robot navigation via intrinsic evolution") by (Tan, K. C., Wang, L. F. and Lee, T. H.). Reproduced by Taylor & Francis Group, LLC., http://www.taylorandfrancis.com

[2000], the robots in next generation should be able to interact with humans and carry out work in unstructured environments. These types of intelligent autonomous systems entice an entirely new set of applications which will become more needed in the years to come in industrialized nations. At the same time great interests have been shown in the development of evolvable hardware (EHW) [Sanchez et al., 1996; Kitano, 1996; Manderick and Higuchi, 1996]: a new set of integrated circuits that are capable of adapting their hardware autonomously to dynamic changes in unstructured environments in real-time. Hardware evolution is an alternative form of system conception that dispenses with conventional hardware design methodology in solving complex problems in a variety of application areas, ranging from pattern recognition [Iwata et al., 1996; Higuchi et al., 1997] to autonomous robotics [Thompson, 1995; Keymeulen et al., 1997; Haddow and Tufte, 1999].

Evolvable hardware has been known recently as an umbrella term to characterize the applications of evolutionary techniques for hardware design and synthesis. This area of research well combines aspects of evolutionary computation with hardware design and synthesis. There is some debate over precisely what the term should encompass. Some researchers believe that it is a learning technique that applies only to the circuit synthesis and not the combinatorial placement and routing stages of the hardware synthesis lifecycle. Others believe that the term should be reserved for hardware that autonomously reconfigures itself whilst in use, according to changes in its current environment. In our view, evolvable hardware should be taken to mean a hardware system that uses an evolutionary algorithm to guide alterations to its architecture.

What constitutes altering hardware architecture is still a gray area. Any alteration of a circuit's behavior can be considered as altering the circuit itself in a loose sense. For example, consider the behavior of a register. It can be imagined as a combinational circuit. Changing the value in the register would then appear as if the combinational circuit had been altered. In this spirit, Salami et al. [2] have developed an evolvable hardware compression architecture where the only part of the architecture under evolutionary control is a value contained in a register. However, Thompson [3] noted that if we take such a definition for evolvable hardware, even a conventional register microprocessor can be though of as an evolvable hardware, as we program the values of its registers to modify its behavior.

This chapter reports our research in constructing an EHW-based autonomous mobile robot navigation system which is capable of realizing the behaviors of light source following and obstacle avoidance in real-time. The robot controller is implemented on a standard autonomous miniature robot, e.g., the Swiss Khepera mobile robot platform [Mondada, et al., 1993]. It is equipped with eight infrared proximity sensors and two DC motors independently controlled by a PID controller. In order to achieve autonomy, the Khepera robot must make its own navigation decision in a complex and dynamic environment, and the artificial evolution is operated on reconfigurable electronic circuits to produce an efficient and powerful evolvable control system for the autonomous mobile robot.

The reconfigurable electronic circuits of Xilinx 6216 Field Programmable Gate Array (FPGA) turret [Xilinx Inc., 1996; Shiratsuchi, 1996; Porter et al., 1999] is chosen as the core module in this work, which can be attached onto the extension ports of Khepera robot via K-Bus. The FPGA can be viewed as a multi-input multi-output digital device with a large number of Boolean functions arranged in a complex way. During the navigation, the input variables defined by the infrared sensors on the robot are fed into the FPGA and the output variables from the device are encoded into different motor speeds. The multi-input multi-output function implemented on the FPGA then provides the desired reactive behaviors in response to any changes in the environments. Therefore the autonomous navigation can be viewed as a real-time optimization process aiming to achieve the best system configuration, i.e., finding the optimal Boolean function controller as the world state changes.

The concept of intrinsic EHW [Thompson et al., 1999] is applied in this study to manipulate the hardware configuration in associated with the world states, which uses Evolutionary Algorithm (EA) [Fogel, 2000; Tan et al., 2001a] to search for bit strings describing circuits that induce a target behavior in the FPGA. The approach of off-line evolution is adopted here for evolving the autonomous robotic system, i.e., the adaptation happens during the learning phase of EHW instead of in an execution mode. The concept of intrinsic and off-line evolution will be elaborated later in the chapter.

The chapter is organized as follows: Section 3.2 introduces the important features of EHW and explains the design strategy of the EHW-based robot navigation system. Section 3.3 looks into the issues of hardware configuration and development platform of the evolvable autonomous system, which include the description of Khepera miniature robot and XC6216 FPGA in the Khepera add-on turret. Section 3.4 presents the physical implementation of the evolvable robot navigation behaviors on light source following and obstacle avoidance with an external robot failure. Conclusions and summary of the findings are drawn in Section 3.5.

3.2   Classifying Evolvable Hardware

Evolvable hardware is a rapidly growing field. It is useful to be able to classify the explosion of new implementations that are constantly appearing. Therefore, a number of features by which evolvable hardware can be classified are discussed here.

3.2.1   *Evolutionary Algorithm*

In the first place, a system should use an evolutionary algorithm to alter its hardware configuration in order to be evolvable. Hence evolvable hardware can be classified according to the evolutionary algorithm used in the application. The most commonly used is the Genetic Algorithm (GA), even though genetic programming has also been used. Other evolutionary algorithms could be used, but there have been no reports of their use in the past literature. Other bio-inspired hardware, for instance, the hardware neural network chips, cannot be considered evolvable hardware unless an evolutionary algorithm is used in their design or operation. Stoica [4] has used a genetic algorithm to evolve neural networks on the JPL's NN-64 chip. Consequently, this can be considered as evolvable hardware.

3.2.2   *Evaluation Implementation*

As with all evolutionary systems, fitness values to evaluate the system performance in the evolutionary algorithm must be calculated for each member of the population in every generation. For this purpose, initial experiments in evolvable hardware used simulations of the hardware that each member of the population specified. This was because the production of hardware was impossible within the time-scale needed to use it for the evaluation stage of an evolutionary algorithm. Evolvable hardware with simulated evaluation was labeled extrinsic by de Garis [5], who also predicted that developments in reconfigurable hardware technology could lead to the possibility of implementing solutions fast enough to evaluate real hardware within an evolutionary algorithm framework. This is labeled intrinsic evolvable hardware. As technology has developed in the past decade, such reconfigurable hardware is now available, most commonly in the form of modern Field Programmable Gate Arrays, or FPGAs.

In the context of field programmable gate arrays, the above evaluation methods can be further appreciated from the following point of view. In the extrinsic evaluations, an evolutionary algorithm produces a configuration based on the performance of a software simulation of the reconfigurable hardware. The final configuration is then downloaded onto the real hardware in a separate implementation step: a quite useful approach if the hardware is only capable of being reconfigured a small number of times. In the intrinsic evaluations, each time when the evolutionary algorithm generates a new variant configuration, it is used to configure the real hardware, which is then evaluated at its task. As a result, an implemented system is evolved directly; the constraints imposed by the hardware are satisfied automatically, and all of its detailed characteristics can be brought to bear on the control problem. Since FPGA is not constrained by small reconfiguration iterations, the intrinsic evaluation is more suitable for such reconfigurable hardware.

FPGAs are VLSI arrays of digital multifunctional blocks connected by routing wires. Both the function of the block and the routing between the blocks are user-defined by programming an on-chip memory. Some FPGAs allow direct configuration of all or part of this memory on a microsecond time-scale, allowing intrinsically evaluated evolvable hardware to be realized, as demonstrated by Thompson. This often allows evaluation many orders of magnitude faster than simulation. Therefore, much more complex problems can be tackled. Intrinsic evaluation also opens up a range of hardware behaviors to evolution that are not simulated in standard hardware simulations, such as pattern recognition, data compression and medical applications.

On the other hand, extrinsic evolvable hardware has two useful features not provided by intrinsic evolvable hardware. Firstly, reconfigurable logic does not lend itself to implementations of pure analogue design as the devices were intended to be used in digital systems. Work with intrinsic evolvable hardware is therefore heavily biased towards digital behavior, which in many circumstances is not a good thing. Extrinsic evolvable hardware can work with any design paradigm, and so any bias towards a particular kind of circuit behavior is limited only by the researcher's imagination and the quality of the simulation used for evaluation. Analogue circuits have successfully been evolved extrinsically in many cases.

Secondly, circuit designs evolved extrinsically can be simulated with only a net list description of the circuit. This avoids any potentially computationally expensive technology mapping stage needed to implement a circuit from a more general genotype representation. However, the time saved comes at the expense of the speed of solution evaluation. This means that only relatively simple circuits evolved using more abstract, lower complexity simulations are likely to gain from this advantage.

3.2.3 *Genotype Representation*

Artificial evolution usually evolves an abstract representation of a system. Hirst [6] identified a number of stages within the hardware design lifecycle where the problem representation could be used as a genotype for an evolutionary algorithm. These range from high-level behavioral descriptions of the hardware, through evolution of Boolean functions to device specific net lists.

As discussed above, we would expect different hardware behaviors to be exploited by different abstractions. Hence possibly very different fitness landscapes may exist for a problem represented by two different abstractions. If a representation from earlier in the design/synthesis lifecycle is used, more details of the target hardware have been abstracted away. This is usually done to create a lower dimension, less modal, landscape, which may be easier for evolution to search. However, there may be points in the solution space that the search space of the abstract genotype representation is not able to reach. Also it is worth to note that landscapes more suitable for search by evolution will only arise if well-designed abstractions are used.

The genetic representation is often deliberately constrained even when using fairly low-level device-specific intrinsically evaluated circuits. For example, they can be forced to behave digitally by imposing phase restrictions on any cell feedback, for instance through a clocked register. Another example of restricted representation with device specific intrinsic evaluation is Miller's combinational representation for Xilinx 6200 FPGAs that does not allow feedback of any kind for the same reasons. This system also restricts the routing distance between cells to reduce the size of the search space. The effects of imposing such constraints are discussed in the next section.

Thompson has successfully used completely unconstrained evolution using the configuration bits of an FPGA as his representation in his tone discriminator. He generated an extremely compact circuit that exhibited behavior so unusual that it still has not been fully analyzed. It should be noted that the search space for this kind of representation is inherently large. Thompson used 1800 bits long chromosome in his evolvable module. Thompson points out that the size of the search space is often not as important as how "evolvable" it is - i.e. how efficiently the shape of the landscape guides the algorithm to areas of high fitness.

3.3   Advantages of Evolvable Hardware

There are a number of features of evolvable hardware that give the technique some advantages over standard hardware design techniques and standard evolutionary techniques. These new features are discussed as follows.

3.3.1   *Novel Hardware Designs*

All circuits contain a large number of components. Human designers need to reduce the search space of all functions of these components to a manageable size. To do this, they tend to work in a space of lower dimensionality in which they are expert. For instance, some designers treat all components as perfect digital gates, when in fact the components are high gain analogue devices. The evolutionary approach may allow us to define the search space in a way that is more natural to both the problem and the implementation, so exploration of designs from the much larger and richer solution space beyond the realms of the traditional hardware search spaces is possible.

- Novelty through Implementation: The first class of novel features that we can take advantage of are those specific to a particular hardware implementation. Conventional designers work with the abstract models of an implementation, or more general device-independent models. Intrinsic evolvable hardware works directly with the hardware implementation, and can take advantage of any features the device offers over and above those in standard hardware models.

  For instance, probably the most extreme form of hardware implementation abstraction is that of the digital design domain. In order for the digital design abstraction to be realized in hardware, we have to restrict the temporal behavior of the physical circuit, imposing set-up and hold times on sections of the circuit. This allows the analogue circuit to appear to behave as digital gates. Restrictions on timing such as these not only prevent us from taking advantage of a huge amount of potentially useful gate behavior, but also force us to spend a great deal of the design effort ensuring that the phase rules are adhered to throughout the circuit. Analogue designers are not immune to the process of implementation abstraction and modeling. They treat all components as perfect analogue devices, trying to avoid phenomena such as parasitic capacitances and electromagnetic coupling, and very rarely taking into consideration even quantum mechanical approximations of how the components behave, let alone their true physical behavior. As we move to smaller fabrication processes, these effects are becoming increasingly more important, and variations between nominally identical hardware will become more apparent. The objective nature of evolutionary systems mean that such effects can be used to our advantage, rather than fought with to make our abstractions work, allowing more efficient circuits with higher yields. On the other hand, such circuits still suffer from bias towards digital behavior, as FPGAs are designed with this behavior in mind. The effects of such a bias should be considered carefully before an implementation.

- Novelty through the Environment: The second class of novel design features arises from real-time interactions between the particular hardware implementation of an evolvable hardware system and its environment, which by definition cannot be simulated perfectly. This class of behaviors is particularly important for real-time systems such as industrial or robot controllers, and is different from the normally non-simulated dynamic interactions between individual hardware components within the system, which were discussed above.

- Novelty through Behavioral Specification: A major distinction between conventional and evolutionary design is that in the latter case the designer must specify the behavior of the circuit, not the architecture. This is the source of the third class of novel behaviors, as the problem is expressed in a less restrictive manner. In addition to modeling component and environmental behavior poorly, humans tend to use a restrictive top-down 'divide and conquer' approach to understand the problem through the use of modules. These are normally defined with respect to space or time. The clear benefit that modularization brings with regard to functionality is to circuits with strict global timing rules, for instance digital circuits. In these circuits, tasks must be completed within a certain time, usually a clock cycle. This means that signal lengths are limited by the width of the clock cycle, or vice versa. Modularization allows the timing constraint to be local each module, allowing these circuits to operate faster, as their critical paths are shorter.

With the removal of circuit-wide phase governance that evolutionary techniques allow, any real restriction to spatial modules is also removed. The Thompson's asynchronous oscillator displayed no clear cohesive networks, so it could be described as exhibiting behavior from this class.

Even without phase, evolution may still find modules useful, but modules defined in dimensions other than space and time may be more natural for a given problem, and so lead to better solutions. Artificial evolution is well suited to exploring these areas of space. But as evolutionary design uses a bottom-up paradigm, we may see complex circuits evolve with no discernible modules, which as a result, are outside the search space of conventional hardware engineers.

None of these three kinds of behaviors are attainable through traditional software implementations of evolutionary algorithms, and so we can expect to see much more economical circuits than would be needed to provide a hardware evolutionary system of similar power. A lack of robustness is characteristic of circuits developed in this way as they may take advantage of features outside the specification of the technology that they are implemented on, hence behavior may not be reproduced in other apparently identical implementations. Lack of robustness to vendor specified environmental conditions might also be evident in evolved hardware of this kind.

3.3.2   *Low Cost*

Evolution can provide a way to reduce the input needed from a hardware designer, hence reduce the final cost of the design. It is important to note that moving the designer's job from hardware design to fitness function design does not necessarily infer that the job is easier, or quicker to complete. However, it is highly expected that evolving hardware for which the fitness functions are easy to design should yield cost savings over standard design techniques. Low cost reconfigurable hardware is often used to embody evolved designs, which further reduces the cost for low volume designs by avoiding the need for a VLSI fabrication process. Utilization of reconfigurable hardware also allows changes in specification to be applied not only to new applications of a design, but also to examples already in use, thus avoiding hardware replacement costs. Risk, with its associated cost, is also reduced since design faults can be corrected, either by hand or through further evolution.

3.3.3   *Speed of Execution*

Hardware implementations of any software system can provide an advantage in terms of speed of execution. This allows evolutionary systems to be applied to many areas requiring real-time responses at a cost level that is unattainable at present stage with traditional software evolutionary implementations. Evolvable hardware has a speed advantage not only for the execution of the final solution, but also for the genetic learning process which involves the evaluation of many trial solutions by their repeated execution. Use of hardware for this part of the evolutionary algorithm can result in a speedup of many orders of magnitude for in genetic learning. This speedup means that evolvable hardware can be used for real-time learning applications that were previously irresolvable with software-based evolutionary algorithms. As solution evaluation is normally the bottleneck of evolutionary systems, many evolvable hardware systems still use conventional software evolutionary algorithms to carry out the genetic operations, whilst using software-reconfigurable hardware for solution evaluation.

3.3.4 *Economy of Resources*

Evolvable hardware systems that are implemented on a single silicon die can be applied to many areas where resources, for instance the area, power or mass of the solution, are limited. Again, this gives an advantage over more resource-hungry software evolutionary systems. Resource constraints can also be included in the evolutionary specification, allowing exploitation of limited hardware resources to be explored much more thoroughly than conventional hardware design methodologies afford.

3.4   EHW-Based Robotic Controller Design

There are two approaches to achieve evolvable adaptive system like autonomous robot navigation in EHW: (1) evolving the system in a noisy environment in which examples of each event are likely to occur [Seth, 1998]; (2) performing the evolution during the operational phrase so that the system is in a continual state of improvement [Yao and Higuchi, 1999]. Both the methods are, however, fraught with problems that are difficult to solve. The first method has the ability of exhibiting certain adaptive behaviors but could only meet up to certain extent the challenges that are introduced during the evolution. The second method requires the chance of new behaviors to be evolved and evaluated. In this approach, not only the evaluation mechanism is unclear, but a system of which continual operation is expected would simply not be able to stop so that new behaviors could be tried out. Therefore it is extremely hard for EHW to achieve adaptation in the operational phase as adopted in the second approach.

A few approaches have been proposed to meet the challenges faced in evolving adaptive systems, and the experiments have shown that a reliable transition from simulation to reality is possible by modeling only a small subset of the agent/environmental properties [Jakobi, 1997; 1998; Jakobi and Quinn, 1998; Husbands, 1998; Smith, 1998]. Grefenstette [1996] stated that while operating in a real environment, an agent should maintain a simulated environmental model which can be updated whenever new features occur either in the environment or the agent's own sensory capabilities, and the learning continues throughout the operational phase. The works of Thompson [1995] and Keymeulen et al., [1997] have shown that the transfer of behavior from simulation to reality is feasible without the loss of robotic behaviors, although some issues still remain open.

3.4.1 *Evolvable Hardware*

This section describes the basic principle of evolvable hardware. Artificial evolution and reconfigurable hardware device are the two basic elements in EHW. According to its application domains, EHW can be applied to circuit and adaptive system designs [Yao and Higuchi, 1999]. There are two views towards what EHW actually is along this line. One view regards EHW as the application of evolutionary techniques to circuit synthesis [de Garis, 1996]. This definition describes EHW as an alternative to conventional specification-based electronic circuit design. Another view regards EHW as the hardware capable of on-line adaptation through reconfiguring its architectures dynamically and automatically [Higuchi and Kajihara, 1999]. This definition specifies EHW as an adaptive mechanism. In the chapter, the essential characteristics in both definitions of EHW, e.g., unlimited hardware reconfiguration times are utilized.

Evolvable hardware is based on the idea of combining reconfigurable hardware devices with evolutionary computing to execute reconfiguration autonomously. The basic concept behinds the combination of these two elements is to regard the configuration bits for reconfigurable hardware devices as chromosomes for evolutionary algorithms [Higuchi et al., 1999]. Fig. 2.1 in the last chapter illustrates the general evolution process in an evolvable hardware. Typically, a population of individuals in the initial population is randomly generated. Using a user-defined fitness function (the desired hardware performance), the EA selects promising individuals in the population to reproduce the offspring for next generation based upon the Darwinian principle of survival-of-the-fittest [Tan et al., 2001b]. If the fitness function is properly designed for a specific task, then the EA can autonomously find the best hardware configuration in terms of architecture bits to realize the desired task.

Each level of hierarchy (unit cells, 4×4 cell blocks, 16×16 cell blocks, 64×64, etc.) has its own associated routing resources. Basic cells can route across themselves to connect to their nearest neighbors and thus provide wires of length 1 cell. Cells used for interconnect in this manner can still be used to provide a logic function. Wires of length 4 cells are provided to allow 4×4 cell blocks to route across themselves without using unit cell resources. Similarly, 16×16 cell blocks provide additional wires of length 16 cells and the 64×64 array provides Chip-Length wires. Larger XC6200 products extend this process to 256×256 cell blocks and so on, scaling by a factor of 4 at each hierarchical level as required. Intermediate array sizes (e.g. 96×96) are created by adding more 16×16 blocks. Those switches at the edge of the blocks provide for connections between the various levels of interconnect at the same position in the array, for example, connecting length 4 wires to neighbor wires.

The longer wires provided at each hierarchical level are termed 'Fast-LANEs'. It is convenient to visualize the structure in three dimensions with routing at each hierarchical level being conceptually above that in lower hierarchical levels, with the cellular array as the base layer. The length-4 FastLANEs are driven by special routing multiplexers within the cells at 4×4 block boundaries. All the routing wires in the chip are directional. They are always labeled according to the signal travel direction.

The benefit of the additional wiring resources provided at each level of the hierarchy is that wiring delays in the XC6216 architecture scale logarithmically with distance in cell units rather than linearly as is the case with the first generation neighbor inter-connect architectures. Since 4×4 cell block boundaries lie on unit cell boundaries, the con-switching function provided at 4×4 cell boundaries is a superset of that provided at unit cell boundaries. It means that it provides for neighbor interconnect between the adjacent cells as well as additional switching options using the length 4 wires. Similarly, the switching unit on 16×16 cell block boundaries provides a superset of the permutations available from that on the 4×4 cell block boundaries. Further switching units are also provided on the 64×64 cell boundaries to provide the Chip-Length FastLANEs.

The XC6216 architecture also provides the global wires and magic wires to form the ample routing resources within the structure to minimize the likelihood of timing problems and provide high-performance bus-turning.

This part of the architecture is not used in our robotic design, therefore, being ignored in this part of explanation.

3.4.2 *Function Unit*

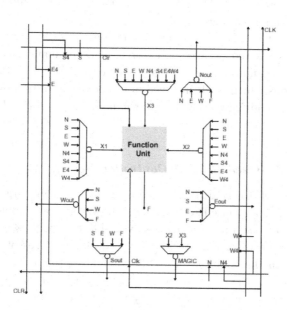

Fig. 3.1 Internal structure of XC6216 basic cell [Xilinx, 1996].

Figure 3.1 above shows the basic XC6200 cell in detail. The inputs from neighboring cells are labeled N, S, E, W and those from length 4 wires N4, S4, E4, W4 according to their signal direction. Additional inputs include Clock and Asynchronous Clear for the Function Unit D-type register. The output from the cell function unit, which implements the gates and registers required by the user's design, is labeled F. The Magic output is used for routing as described earlier. The multiplexers within the cell are controlled by bits within the configuration memory.

A feature of the XC6200 architecture is that a block within the user's design can be dynamically replaced with another block by the host processor, reconfiguring only the corresponding area of the control store. In our experiment, such block replacements are necessary in the crossover and mutation operations. All the configuration memory can be accessed as 8-bit bytes.

In particular, only first three address modes are used in our design. In address mode 00, each cell has 3 separate 8-bit configuration bytes and 1 single-bit state register, which configure the internal cell structure and the connections with neighboring cells. The address mode 01 and 10, handle the connections between the edge cells and external data pins, which only need to be configured once at initialization.

The concept of a Boolean function controller defined within the FPGA device can be viewed as a combinatorial block of many basic cells. To evolve the Boolean function that maps a set of input states into the desired output motions, all the cells involved in the design need to be reconfigured iteratively by modifying every bits in the mode 00 until the desired result is obtained.

3.4.3   *EHW-Based Robotic Controller Design*

A wide range of artificial intelligence techniques have emerged for building autonomous and intelligent systems. However, these implementations are often costly as compared to the evolvable hardware, which makes it possible to use evolutionary techniques to evolve patterns of behavior for the robot in the hardware level by fully exploiting the hardware resources and dynamics. In this section, various design considerations and strategies in EHW-based autonomous robotic systems are described, which include the topics of Boolean function controller, chromosome representation, extrinsic and intrinsic evolution, off-line and on-line adaptation, and robot navigation tasks.

As mentioned earlier, Genetic Algorithm (GA) is the driving force behind our evolvable hardware. In principle, GA is a simple iterative procedure that consists of a constant-size population of individuals, each one represented by a finite string of symbols, known as the genome, encoding a possible solution in a given problem space. This space, referred to as the search space, comprises all possible solutions to the problem at hand. Generally, the genetic algorithm is applied to space which is too large to be exhaustively searched. The symbol alphabet used is often binary, though other representations have also been used, including character-based encodings, real-valued encodings, and – most notably – tree representations. As can be seen later, our evolvable hardware adopts the ordinary binary representation simply because the FPGA configuration bits are represented in the binary form.

The standard genetic algorithm proceeds as follows: an initial population of individuals is generated at random or heuristically. Every evolutionary step, known as a generation, the individuals in the current population are decoded and evaluated according to some predefined quality criterion, referred to as the fitness, or fitness function. To form a new population (the next generation), individuals are selected according to their fitness. Many selection procedures are currently in use, one of the simplest being Holland's original fitness-proportionate selection, where individuals are selected with a probability proportional to their relative fitness. This ensures that the expected number of times an individual is chosen is approximately proportional to its relative performance in the population. Thus, the high-fitness individuals stand a better chance of "reproducing", while the low-fitness ones are more likely to disappear.

Selection alone cannot introduce any new individuals into the population, i.e., it cannot find new points in the search space. These are created by genetically inspired operators, of which the most well known are crossover and mutation. Crossover is performed with probability p_{cross} between two selected individuals, called parents, by exchanging parts of their genomes to form two new individuals, called offspring. In its simplest form, substrings are exchanged after a randomly selected crossover point. This operator tends to enable the evolutionary process to move toward "promising" regions of the search space. The mutation operator is introduced to prevent premature convergence to local optima by randomly sampling new points in the search space. It is carried out by flipping bits at random, with some probability p_{mut}. Generally, genetic algorithms are stochastic iterative processes that are not guaranteed to converge. The termination condition may be specified as some fixed, maximal number of generations or as the attainment of an acceptable fitness level. In our application, the termination condition is relatively relaxed and specified as the attainment of the second highest fitness score. Below presents the standard genetic algorithm in pseudo-code format.

begin GA
    g:=0 {generation counter}
    Initialize population P(g)
    Evaluate population P(g) {i.e., compute fitness values}
    while not done **do**
        g:=g+1
        Select P(g) from P(g-1)
        Crossover P(g)

        Mutate P(g)
        Evaluate P(g)
    end while
end GA

3.4.3.1 *Boolean function controller*

It is assumed that the robotic behavior can be described by a Boolean function which is described by a function F of n Boolean variables that maps sensor inputs to motor outputs according to the desired sensory-motor response. There is an adequate means of encoding the function F for applying the evolutionary computing techniques. The Boolean function is extremely suitable since it is readily expressed in binary codes. A population of potential functions F_s expressed in bit strings can be generated and operated on by evolutionary algorithms to determine the behaviors of a robot. It is able to perform the pre-specified robotic tasks in a reactive manner as world state changes and is well-suited for an evolutionary search algorithm. The target of the Boolean function approach is to find out a function F of n Boolean variables that represents the desired reactive behaviors. The input domain of the function F is $\{0, 1\}^n$ where 2^n is the possible world states detected by the robot and the output domain of the function F is $\{0, 1\}^m$ where 2^m is the possible motions. In such a context, the input domain is encoded into 4 Boolean variables $x_i (i = 1, 2, 3, 4)$, each corresponding to two neighboring IR sensors. Detection by either adjacent sensor sets the corresponding Boolean variable. Hence, there are totally 16 world states detected by the robot. Each point of the input space represents a robot situation and the associated output $y = F(x_0, x_1, \ldots, x_{n-1})$ represents the motion it will perform in the situation. The goal is to identify the unknown function from a given set of observable input-output pairs and a fitness score that represents whether and how well the unknown function produces the expected behavior.

    As depicted in Fig. 3.2, the adaptation task is to find an appropriate function F, mapping 16 inputs (world states) to 4 outputs (motions), in a search space of 416 functions from a set of observable input-output pairs. The evolutionary algorithm performs a parallel search in the space of Boolean functions in order to find the optimal Boolean function that exhibits high performance in FPGA. The algorithm is implemented in hardware, where the 121 bits of the EHW are regarded as a chromosome in the evolutionary algorithm. It generates new Boolean functions by transforming and combining the existing functions using conjunction, disjunction and negation operators in a genetically inspired way. Statistics related to the performance of the functions in the hardware level are also used in a genetic way to guide the search as well as to select appropriate functions to be transformed and combined.

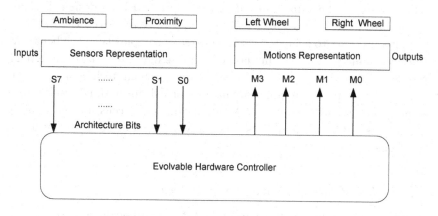

Fig. 3.2   EHW-based robotic controller.

3.4.3.2   *Chromosome representation*

The basic idea of EHW is to regard the architecture bits of the FPGA as a chromosome in an evolutionary algorithm. If these bits change, the architecture and the function of the logic device are also changed. Therefore, the basic idea is to let the architecture bits evolve in order to adapt the function of the logic devices to perform the desired behavior. The EA performs a parallel search, in the space of Boolean functions, for a set of functions that exhibits the desired behaviors in the hardware. All features of the FPGA (cell functions and connections) can be accessed by a genotype for exploitation by the evolutionary process. A genotype used on the turret consists of genes, each of which specifies how a cell should behave in the evolution. A gene consists of 3 bytes (i.e., 24 data bits) of information specifying the cell functions and connections between the cell's neighboring cells or the FPGA's input/output lines. All of a cell's 40 functions can be selected, and all 3 modes of cell connections (Fast LANEs, Length-4 Fast LANEs, Length-16 Fast LANEs) in all 4 directions (north, south, east, and west) can be specified. The genotypes, expressed as a string of bytes at the host and robot microprocessor, are converted into a bit string on XC6216 by hardware for fast loading.

In general, tens of cells are needed to form a complicated Boolean function to meet the requirement of evolving a desired behavior such as light source following. Hence, a single chromosome contains $24 \times N$ configuration bits, where N is the number of cells exploited in the FPGA design. It is important to ensure that a sufficiently large area of resources on the FPGA is exploited in order to achieve a certain degree of complexity in the behavior. However, unnecessary or redundant cells in use may cause the evolutionary process to slow down. Therefore the network growing technique has been applied here to find an appropriate number of cells for optimal results. The technique starts with a sufficiently small number of cells for accomplishing the task at hand, and adds a new row or column of cells only if the requirements are not met.

3.4.3.3 *Evolution and adaptation methodology*

In this section, the evolution and adaptation methodologies used in the study are presented. According to the number of hardware configurations at each generation during the evolution, the fitness evaluation strategy along the dimensions of extrinsic and intrinsic evolution is first discussed. The approaches of off-line and on-line adaptation are then discussed according to the phase where the evolution takes place.

A. Extrinsic and intrinsic evolution

Much of the current research on evolutionary autonomous agents is centered around simulations of neuro-mimetic networks performed on a general-purpose computer. In the context of hardware evolution, this method should be classified into extrinsic evolution. The extrinsic evolution makes the assumption that simulated characteristics of the hardware are carefully designed, and the evolvable Boolean controllers are robust enough to deal with any uncertainties or inaccuracies of the motor responses and sensor values. Obviously, the method of extrinsic evolution is not suitable for our applications, since it is very difficult to simulate all the hardware characteristics, e.g., hardware abstraction often makes full exploitation of the hardware resources and dynamics infeasible.

Intrinsic evolution is adopted in this study, e.g., evolutionary process takes place on the microprocessor of Khepera robot instead of the host computer. Each genotype in the population describes functions and connections of the cells of FPGA as well as other parameters that define the reconfigurable hardware. Once a new generation of individuals is established via genetic recombination, they will be implemented, one genotype at a time, on board the FPGA with the help of Khepera's microprocessor. The performance of the FPGA (phenotype) is then evaluated by the evaluation mechanism on the robot microprocessor. The genotype also dictates the selection of cell functions and inter-cell connections.

In intrinsic hardware evolution, FPGA's physical characteristics are fully explored to achieve the desirable sensory-motor responses. The objective is to extract from whatever physical characteristics of the configured FPGA desirable input (sensor) - output (motor) signal relationships towards achieving the fitness criteria. Many of the causes which are responsible for the evolved signal relationships (mappings from inputs to outputs of the FPGA) may not be readily apparent. However, the intrinsic evolutionary process actively exploits side effect and obscure physical phenomena of the evolved hardware to improve the fitness value. The Khepera's on-board processor mediates the transfer of the genetic information to and from the FPGA. The real-time physical implementation results reported later show that intrinsic hardware evolution is powerful and capable of exploiting hardware resources by using a surprising small portion of the FPGA cells.

B. Off-line and on-line adaptation

Most of the research in EHW adaptation is based on off-line evolution, which is also adopted in this work. In the approach, the adaptation phase precedes the execution phase, i.e., the adaptation happens during the learning phase of EHW instead of in an execution mode. One merit of using the off-line hardware evolution is that it allows the preliminary studies of evolutionary process prior to the real-time implementation [Nolfi and Floreano, 2000]. The physical results reported later in the chapter show that with the implementation of off-line learning and intrinsic EHW, the number of learning and computation time needed to derive the evolvable robotic controller is considerably reduced and the hardware resources are exploited to the full.

The feasibility of on-line evolution was also explored in our study for evolving robot navigation behaviors, which will be elaborated later in detailed in Section 3.1. The implementation results obtained were unsatisfactory and the evolved navigation behaviors were found to be inconsistent. Due to the trial-and-error nature of EAs, the poor individuals could cause severe damages to EHW or the physical environment in which it is being evaluated, if there is no additional technique to prevent them from happening [Yao and Higuchi, 1999]. Moreover, the population-based evolutionary learning used by all EHW at present cannot make the learning to be incremental and responsive. Besides the drawback of computationally expensive due to its real-time interactions with environments, on-line adaptation also faces the problem of evaluating fitness function. For example, a mobile robot often needs to decide the motor speeds based on real sensory information in on-line robotic navigation. However, different moves will generate different sensory information at the next time step, and this information will be used to evaluate the fitness for the corresponding move. Therefore on-line EHW adaptation is very difficult to achieve, particularly in the application of autonomous robotics.

3.4.3.4 *Robot navigation tasks*

Two physical robotic navigation experiments are adopted in this chapter to validate the proposed EHW-based autonomous robotic system. The task of the robot in the first experiment is to follow a moving torchlight, e.g., a torch held by the experimenter will act as the light source in the experiment. In the second experiment, the robot is required to exhibit perfect obstacle avoiding behavior using a robot with traction fault, e.g., one of the robot wheels is out of control and can only keep a constant speed. The details on both experiments will be given in Section 3.4.

3.5 Hardware and Development Platform

3.5.1 *Sensor Information*

The geometrical shape and the motor layout of Khepera provide for easier negotiation of corners and obstacles when its control system is still immature. Its small size and good mechanical design provide intrinsic robustness. In its basic version it is provided with eight infrared proximity sensors placed around its body that are based on emission and reception of infrared light. The details of the infrared sensors will be elaborated in this subsection. Another advantage possessed by Khepera robot is that several new single sensors and complete modules such as an FPGA module and a gripper module can be easily added, thanks to the hardware and software modularity of the system.

The Khepera robot is provided with eight infrared proximity sensors placed around its body (six on one side and two in the opposite side), which are based on emission and reception of infrared light. They are placed around the robot and are positioned and numbered as shown in Fig. 3.3. These sensors embed an infrared light emitter and a receiver. This sensor device allows two measures. Each receptor can measure both the ambient infrared light, which is rough measure of the local ambient light intensity in normal conditions, and the reflected infrared light emitted by the robot itself (for objects closer than 4-5 cm in our experiment). These measures do not have linear characteristics, are not filtered by correction mechanisms, and depend on a number of external factors, such as the surface properties of objects and the illumination conditions.

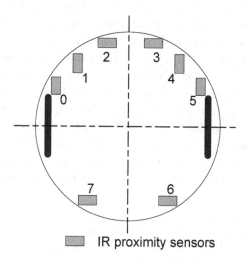

Fig. 3.3   Sensory-motor layout.

- The normal ambient light. This measure is made using only the receiver part of the device, without emitting light with the emitter. A new measurement is made every 20 ms. During the 20 ms, the sensors are read in a sequential way every 2.5 ms. The value returned at a given time is the result of the last measurement made. The ambient light was used in the light source following experiment.
- The light reflected by obstacles. This measure is made emitting light using the emitter part of the device. The returned value is the difference between the measurement made emitting light and the light measured without light emission (ambient light). A new measurement is made ev-

ery 20 ms. During the 20 ms, the sensors are read in a sequential way every 2.5 ms. The value returned at a given time is the result of the last measurement made. The output of each measurement is an analogue value converted by a 10 bit A/D converter. The proximity light value was used in the obstacle avoidance experiment. The measured value decreases when the intensity of the light increases. The standard value in the dark is around 450. This feature has been used in our experiment to fulfill the light source following task in the environment. The lowest measured value indicates the direction of the light source (i.e., the lightest direction detected). The typical measurements made on six sensors of the same robot placed in identical conditions are shown in this figure. Small differences of conditions, such as vertical orientation of the sensor, ambient light conditions and color of the floor, can bring additional differences. To reduce light reflection from the floor as much as possible, the standard robot soccer field was chosen to conduct the experiment. The special field has the black coarse color floor and the white glossy wall, minimizing the reflected infrared light from the floor. Since the physical sensor may differ much in their characteristics, adjacent sensors are grouped together to form a logical sensor so that the performance of the sensors is more reliable than treating them individually (from the perspective of fault tolerance, it belongs to the conventional redundancy method).

The means of serial communication is provided for the robot to exchange information with a host computer through the S serial cable. On the host side, the link is made by a RS232 line. The interface module converts the RS232 line into the S serial line available on the robot.

3.5.2 *FPGA Turret*

A Khepera add-on turret is employed for evolutionary experiments in hardware on a single or a group of Khepera robots using FPGA. High performance FPGA, XC6216 by Xilinx which has the following specifications, is used. (1) 4096 cell arrangement in a 64 × 64 matrix format; (2) each cell has some 40 logic and other functions; (3) detailed structure, rich registers, and gates; (4) high speed SRAM control memory; (5) 220 MHz basic clock; (6) 16-bit input and 7-bit output are connected to Khepera through K-BUS. Table 3.1 summaries the general specifications of XC6216 FPGA chip.

Table 3.1   XC6216 FPGA Specifications.

Typical Gate Count Range	# of Cells	# of Registers	# of IOBs	Cell Rows×Columns
16000–24000	4096	4096	256	64×64

3.5.2.1   *Description*

An XC6216 part is composed of a large array of simple, configurable cells [Xilinx Inc., 1996]. Each basic cell contains a computation unit capable of simultaneously implementing one of a set of logic level functions and a routing area through which inter-cell communication can take place. The structure is simple, symmetrical, hierarchical and regular, allowing users to quickly make efficient use of the resources available.

The Xilinx XC6216 FPGA was chosen for our implementation for a number of reasons. (1) It may be configured indefinitely; (2) It may be partially reconfigured; (3) Its design ensured that invalid circuits do not harm it; (4) Its specification is nonproprietary. Point 1 allows its use in repeated fitness evaluation which is the central component of a GA. Point 2 guarantees that configuration time (a nonnegligible cost in fitness computation) is linear in the size of the circuit being loaded rather than proportional to the dimensions of the FPGA. Point 3 tolerates random circuit configurations which are prevalent in the circuit population maintained by the GA. Point 4 permits bit-level circuit description while bypassing conventional circuit entry tools.

3.5.2.2   *Architecture*

The XC6216 architecture may be viewed as a hierarchy. At the lowest level of the hierarchy lies a large array of simple cells. This is the "sea of gates". Each cell is individually programmable to implement a D-type register and a logic function such as multiplexer or gate. A cell can also be configured to implement a purely combinatorial function, with no register. First generation fine-grain architectures implemented only nearest-neighbor interconnection and had no hierarchical routing. XC6200 is a second generation fine-grain architecture, employing a hierarchical cellular array structure. Neighbor connected cells are grouped into blocks of 4×4 cells which themselves form a cellular array, communicating with neighboring 4×4 cell blocks. In the basic XC6216 cell, the inputs from neighboring cells are labeled N, S, E, W and those from length 4 wires N4, S4, E4, W4 according to their signal direction. Additional inputs include Clock and Asynchronous Clear for the Function Unit D-type register. The output from the cell function unit, which implements the gates and registers required by the user's design, is labeled F. The multiplexers within the cell are controlled by bits within the configuration memory. As can be seen from figure, the basic cells in the array have inputs from the length 4 wires associated with 4×4 cell blocks as well as their nearest neighbor cells. The function unit design allows the cells to efficiently support D-type registers with Asynchronous Clear and 2:1 multiplexers, as well as all Boolean functions of two variables (A and B) chosen from the inputs to the cell (N, S, E, W, N4, S4, E4, W4).

3.5.2.3 *Configuration bits*

A feature of the XC6216 architecture is that a block within the user's design can be dynamically replaced with another block by the host processor, reconfiguring only the corresponding area of the control store. In our experiment, such block replacements are necessary in the crossover and mutation operations. The format of the address bus to the XC6216 device is shown in Table 3.2. All the configuration memory can be accessed as 8-bit bytes. The Address Mode bits are used to determine which area of the control store is to be accessed. For example, in Mode 00 the 6-bit row and column addresses are effectively a Cartesian pointer to a particular cell. $(0, 0)$ is the cell in the South-West corner of the array. Once a particular cell has been pin-pointed, the 2-bit Column Offset determines which cell configuration RAM byte is accessed. Each cell has 3 separate 8-bit configuration bytes and 1 single-bit state register.

Table 3.2   Address mode selection.

Mode 1	Mode 0	Area Selected
0	0	Cell Configuration and State
0	1	East/West Switch or IOB
1	0	North/South Switch or IOB
1	1	Device Control Registers

In particular, only first three address modes are used in our design. In address mode 00, each cell has 3 separate 8-bit configuration bytes and 1 single-bit state register, which configure the internal cell structure and the connections with neighboring cells. The address mode 01 and 10, handle the connections between the edge cells and external data pins, which only need to be configured once at initialization.

The concept of a Boolean function controller defined within the FPGA device can be viewed as a combinatorial block of many basic cells. To evolve the Boolean function that maps a set of input states into the desired output motions, all the cells involved in the design need to be reconfigured iteratively by modifying every bits in the mode 00 until the desired result is obtained.

3.5.3 *Hardware Configuration*

As shown in Fig. 3.4, the Khepera miniature robot and the FPGA turret have been selected for implementing the EHW-based autonomous robot navigation system. The Khepera robot is provided with eight infrared proximity sensors placed around its body (six on one side and two in the opposite side), which are based on emission and reception of infrared light. Each receptor can measure both the ambient infrared light, which is rough measure of the local ambient light intensity in normal conditions, and the reflected infrared light emitted by the robot itself (for objects closer than 4-5 cm in our experiment). The ambient light value decreases when the intensity of the light increases. The standard value in the dark is around 450. This feature has been used in our experiment to fulfill the navigation task of seeking and following the light source in the environment, i.e., the lowest measured value indicates the direction of the light source. The reflected infrared light is used to check if there is any obstacle in the proximity of the robot in the anti-collision experiment.

Fig. 3.4   Hardware configuration.

In the dynamic reconfiguration case, the FPGA is used to instantiate different circuits at various stages of the computation. It may be used as a reconfigurable co-processor to a host microprocessor, which is what happens between the FPGA turret and the Khepera base microprocessor. The XC6216 [Xilinx Inc., 1996] used in our experiment is such a chip. Its structure is simple, symmetrical, hierarchical and regular, which allows users to easily make efficient use of the resources available. It consists of an array of hundreds of reconfigurable blocks that can perform a variety of digital logic functions, and a set of wires to which the inputs and outputs of the blocks can be connected. The working principle such as what logic functions are performed by the blocks, and how the wires are connected to the blocks and to each other, can be regarded as being controlled by electronic switches. The settings of these switches are determined by the contents of digital memory cells. For instance, if a block performs any one of the 24 Boolean functions of two inputs, then four bits of this "configuration memory" are needed to determine its behavior. The blocks around the periphery of the array have special configuration switches to control how they are interfaced to the external connections of the chip which are closely associated with the IOB.

A useful feature of the XC6216 architecture is that a block within the user's design can be dynamically replaced with another block by the host processor via reconfiguring only the corresponding area of the control store. In our experiment, such block replacements are necessary in genetic crossover and mutation operations [Goldberg, 1989]. All the configuration memory can be accessed as 8-bit bytes. Two address mode bits are used to determine which area of the control store is to be accessed according to Table 3.2. In particular, only the first three address modes are used in our design. In address mode 00, each cell has 3 separate 8-bit configuration bytes and 1 single-bit state register, which configure the internal cell structure and the connections with neighboring cells. The address modes of 01 and 10 handle the connections between the edge cells and external data pins, which only need to be configured once at the initialization. The concept of Boolean function controller defined within the FPGA device can be viewed as a combinatorial block of many basic cells. To evolve the Boolean function that maps a set of input states into the desired output motions, all cells involved in the design need to be reconfigured iteratively by modifying every bits in the mode 00 until the desired result is obtained.

3.5.4 Development Platform

The mobile robot is equipped with a Motorola 68331 microprocessor which can be connected to a host computer. The microprocessor has 256 KB of RAM and a ROM containing a small operating system [Franzi, 1998]. The operating system has simple multi-tasking capabilities and manages the communication with the host computer. In our study, ANSI C was adopted to develop the program and the source code (i.e., the .c file) was cross-compiled into a downloadable file (i.e., the .37 file, a file format of Motorola Inc.) via C cross-compiler named Cygnus on Windows platform. It is possible to control the robot in two ways [Wang et al., 2000]. As shown in Fig. 3.5, the control system may be run on a host computer with data and commands communicated through a serial line. Alternatively the control system can be cross-compiled on the host computer and downloaded to the robot, which then runs the complete system in a stand-alone fashion. The evolvable hardware system described in this chapter adopts the latter approach.

As shown in Fig. 3.6, the evolution monitoring process is managed by the software on the host computer while all the main processes such as sensor reading, motor control, and genetic operations are performed by the onboard micro-controller. A cable connection between the robot and host computer is established here, which allows the tracking of physical robot behaviors for on-line evolution monitoring and analysis. The evolution information such as generation number, the best chromosome, and convergence trace can be displayed on Hyper Terminal, which is a common terminal emulator on Windows operating system. Such information can be utilized by the designers for parameters adjusting in the EA. The cable is also used to supply electrical power to the robot, e.g., a useful feature for experiments in evolutionary robotics where the robot may spend a long time for evolving the target behaviors. On the robot, the microprocessor acts as an interface between the external sensor-motor system and the field programmable device in which any Boolean function can be implemented. The binary bits from the FPGA output are first read by the controller, which then interprets the bits into the corresponding motion. This configuration allows a complete and precise analysis of the functioning of the robotic control system.

In our experiment, the FPGA add-on turret of Khepera robot is adopted to handle the intrinsic evolution of functions, connections, and the physical characteristics of the cells in FPGA. As described earlier, the FPGA

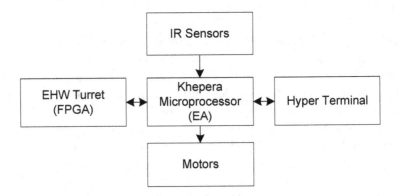

Fig. 3.5   The robotic control system setup.

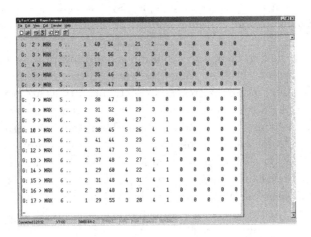

Fig. 3.6   Hyper Terminal displaying evolution information in real-time.

consists of a sea of cells, which can be configured as different logic functions or latches. To configure each cell, three bytes of data are required, resulting in a 24-bit chromosome string. Each bit of the string defines either connection of the cell with its neighboring cell and high-speed lanes or its own function. As shown in Fig. 3.7, it can be downloaded from the host computer to the FPGA on the turret with the help of Khepera's microprocessor. The outputs from the FPGA's 7 output lines are posted on K-bus and read by the robot microprocessor, which are then processed by the robot microprocessor or sent to the host computer for evaluation.

In the case of a small population size, the entire set of genotypes and the mechanism for managing the process of artificial evolution, could be stored on-board the Khepera processor and allows a host free evolution. In our experiment, the evolution process takes place at the robot microprocessor which is capable of coping with all the necessary evolution operations.

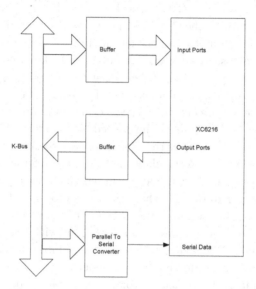

Fig. 3.7   XC6216 K-Bus communication with the robot.

3.6   Implementation Of EHW-Based Robot Navigation

In this section, the configuration of an FPGA is placed under the direct control of unconstrained intrinsic hardware evolution for deriving two interesting robotic tasks. It is found that the evolution solves the problems well, using only a surprisingly small region of the FPGA.

3.6.1   *Preliminary Investigation*

Various methodologies were investigated in the study to find the most suitable approach for our hardware evolution system, including the method of on-line intrinsic evolution. The experiment employed for the evolutionary training consists of a $130 \times 90cm^2$ arena delimited by white painted wooden walls. The floor was made of plywood with smooth surface painted in deep grey. For on-line intrinsic evolution, the arena is filled up with many small white blocks of Styrofoam. Such materials are chosen in the experiment due to its soft nature so that the impact of collision is minimized and the robot is gracefully protected from any violent collision. The obstacle blocks are densely distributed in the enclosed field so that the robot has higher chances to interact with the obstacles and different sensor input states could be presented to the robot in an efficient manner.

For each chromosome, the reflected infrared is measured twice, i.e., one for sensor input states and the other for sensor-based fitness evaluations. Instead of assigning fitness value based on FPGA outputs, the robot moves in the physical environment for 0.3 second and evaluates the chromosome based on the new sensor states. The desired behaviors lead to a state where the robot faces fewer obstacles than in the previous state. The difference in the sensor states is used to calculate the fitness score for each chromosome. However, the on-line fitness evaluation turned out to be impractical because too much time is required for the evaluation process. Therefore on-line evolution considerably decreases the efficiency of the evolutionary approach, due to the need of a huge number of interactions with the environment for learning an effective behavior, e.g., the huge number of interactions with the environment for all individuals = number of generations × number of individuals number of interactions for one individual. In comparison, the genetic operations only take a small amount of time, e.g., the number of genetic operations = number of individuals × number of generations.

The time taken to obtain a satisfactory controller in on-line intrinsic evolution is often dominated by how long it takes to evaluate each variant controller's efficiency (fitness evaluation) for a particular task. It is difficult to evaluate individuals on-line due to the large number of interactions with the physical environments. In particular, since actions to be taken in the current sensor input state affect the future sensor state, it is difficult to present regular sensor input patterns to each chromosome in an efficient manner. If the problem cannot be resolved, the fitness function approach applied to off-line hardware evolution cannot be used in the current situations.

Two physical experiments were conducted to find solutions to the current dilemma. In the first experiment, a random motion was carried out after each fitness evaluation via the FPGA output values (the wheel speeds). Such a random motion may generate a new world state, which is different from previous world states for the current chromosome. In the evolution, six classified input states need to be shown to the robot requesting for its response, and the fitness evaluation is then carried out. Therefore many random motions are needed to search for different input states for each chromosome, and a lot of time is consumed for such an exhaustive state search that causes the evolution to be unpredictable.

The second experiment is similar to the first one in principle, but adopts a different strategy to search for the input states. In this approach, instead of looking for all the defined input states for consecutive chromosomes, algorithm is applied to search for the defined input states for all chromosomes at each generation. Once duplicated input state is presented to the robot, the next chromosome that has not encountered the current input state will take over the current chromosome for fitness evaluation. Such an approach seems to evaluate all the chromosomes in each generation at a much faster speed than the previous method. However, it is still not capable of coping with the massive data needed to process in our application, and a lot of time is wasted in searching for the sensor input states rather than in the intrinsic evolution.

The fitness function used in the off-line evolution is not ideal for the on-line hardware evolution to render the beauty of on-line intrinsic evolvable hardware and to solve the hardware failures mentioned earlier. In the off-line intrinsic evolution, fitness score is solely based on the FPGA output signals in response to the ideal sensor input states and the assumption that sensor-motor system has a reliable interface with the FPGA. This fitness function defined in this way does not guarantee the hardware failures in the sensor-motor system. To place the entire robot system into consideration and let the robot be tolerable to any hardware failure, the fitness function should also possess a mechanism to examine the functionalities of the sensor-motor system. The sensor feedback is considered as a good indicator and the following fitness function has been developed to achieve the above objectives.

For on-line intrinsic evolvable hardware, the evolution process is slightly complicated compared to that of the off-line intrinsic evolvable hardware. The different algorithm has been developed and is shown in Figure 3.8. The reflected infrared are measured twice, the first reading for sensor input states and the second for sensor-based fitness evaluation. Once the real world input states are read by the FPGA, the FPGA internal structure processes the information and then generates a set of output signals that in turn are transferred o motor actions. Instead of assigning a fitness value based on the FPGA outputs, the robot moves in the real world for approx-

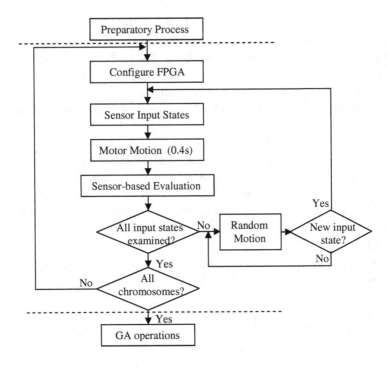

Fig. 3.8   Sensor-based on-line fitness evaluation.

imately 0.3-0.4 second and then evaluates chromosome fitness based on the new sensor states. By right, the desirable behaviors should lead the robot to face fewer obstacles than the previous sensor state. The difference in the sensor states is used to calculate the fitness score of each chromosome. Fig. 3.9 further illustrates the fitness evaluation algorithm.

As shown in Fig. 3.9, different weights are assigned to the 6 front proximity sensors on the robot, and the two back sensors are ignored since we only consider the forward movements of the robot. The closer the sensor is situated to the central division line shown in Fig. 3.9, the higher the weight is assigned to the corresponding sensor. Reason behind such weight assignments is demonstrated in the right portion of Fig. 3.9. When a sensor detects any obstacle, the corresponding sensor weight is added to the total weight for that world state. The most desirable behavior of the robot can be reduced to minimizing the total detected weight in the environment. The initial sensor state weight is compared with the second sensor state weight subject to motor motion and the fitness score is calculated as the

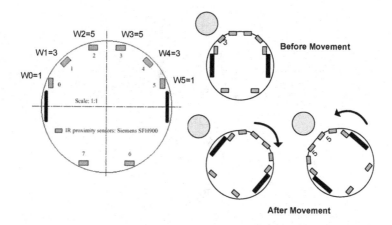

Fig. 3.9   Illustration of sensor-based fitness evaluation.

difference between the two weights. In the above figure, the detected weight for the sensor state prior to the motor movement is only 3. If the robot makes a right turn as a result of the FPGA processing, the robot is safely avoiding the obstacle and the second sensor reading gives a zero weight. On the contrary, the left turn make the situation become worse and the weight increases to 10, which is undesirable according to our on-line fitness function. Suppose that the robot ignore the sensor state and continue to move forward, the weight probably will not change too much. It can be seen that this fitness function produces the same result as what we want to see. However, this strategy is only applicable to the world states in which an obstacle is detected and not appropriate for open environment. Similar to off-line intrinsic evolution, such a state can be ignored since the response in this state is determined by the light source search strategy in the final behavior.

Such an on-line fitness evaluation still turned out to be unsuccessful. Theoretically, it only works in an environment where the obstacles are sparsely situated. However, to speed up the evolution, we placed as many obstacles as possible in the experimental field so that the robot had close contacts with the obstacles and different world states were presented to the robot in a faster pace. If the obstacles are loosely located in the field, the duration of the experiment becomes unpredictable and endless again. Most importantly, one possible condition failed to be considered in our initial thought. The condition is illustrated in Fig. 3.10. In the figure, the robot

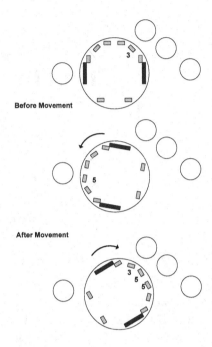

Fig. 3.10   Hidden problem in the on-line fitness evaluation.

should make a left turn in the sensor state prior to the motor movement. However, due to the obstacle on the left unseen by the robot, the desirable movement results in a higher weight, which our fitness function marks as an undesirable motion in this case. This could happen in the environment where the obstacles are densely situated. Furthermore, the period that the motor wheels are allowed to rotate is also arguable. If the rotation period is too short, the effect of the motion is not obvious. If the rotation period is too long, the above situation will take place more frequently and the corresponding results are doubtful.

Several on-line intrinsic evolutions based on the above fitness evaluation have been experimented and no obstacle avoidance behavior has been realized. In summary, the means of fitness evaluation in the above approach is unreliable. As a result, even though some chromosomes have reached the maximum scores, their corresponding FPGA configurations failed to show the desirable results on the robot.

Although the approach of on-line evolution was not satisfactory in this experiment, the implication is far more meaningful, e.g., the intention of constructing an entire sensor-robot system is a better thought than the constrained approach of FPGA alone. In fact, if the simulation of environment could be implemented in a special purpose EHW situated next to the evolved hardware controller [de Garis, 1996], the on-line fitness evaluation should be less time-consuming and more effective. Although some issues remain open, such as the difficulty of simulating interactions between the robotic control system and the environments, it is possible that simulating the environment in a special purposed hardware can become an important tool as new techniques are developed. Moreover, if some sort of bumping sensors could be integrated into the robot and the fitness evaluation is based on results obtained from the bumping sensors, the above approaches should be more reliable.

Therefore, off-line intrinsic hardware evolution was adopted in our work to realize the two robotic behaviors of light source following and obstacle avoidance. As detailed in the following sections, it is found that the off-line intrinsic EHW is capable of fully exploiting the hardware resources in FPGA, and the two robotic behaviors are successfully implemented in physical environments.

3.6.2   Light Source Following Task

In the first experiment, the robot was evolved intrinsically to exhibit behavior of light source following. The experimental diagram is shown in Fig. 3.11. The ambience light emitted by the torch light is detected by the robot sensors and their values are coded into a binary string to be fed into the FPGA. The outputs of the FPGA are used to guide the robot movements.

3.6.2.1   *Software structure of light following task*

As shown in Fig. 3.12, the program of the light source following task comprises of two phases, i.e., the evolution and execution phases. The evolution phase is performed off-line using detailed training events, which is considered to be the essence of intrinsic hardware evolution because every chromosome is used to configure the associated area of the reconfigurable hardware, and the corresponding hardware outputs are used to evaluate the performance of the individuals. The best configuration pattern for the light source following behavior found in the learning or evolution phase is downloaded to the FPGA. The light intensity of the environment is measured by the robot sensors, and the robot keeps on acquiring the set of

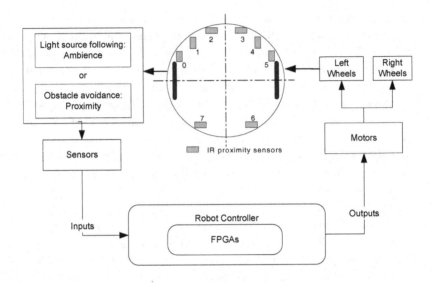

Fig. 3.11   An autonomous robotic system.

sensor values to check if there is any light source around it. If there is no light source detected in its vicinity, the robot will go forward as straight as possible; otherwise the input state is passed onto the FPGA turret to produce the ideal response in following the light source. The approach is efficient in evolving the robotic tasks, since it avoids the on-line interactions with physical environments, which are time consuming and undesirable in this application as discussed in the earlier section.

3.6.2.2   *Program settings of light following task*

The evolutionary parameters such as population size, crossover rate, mutation rate, generation size and etc., were set based upon the evolution speed and fitness score. For example, a population size of 121 was adopted in the experiment after observing that there is little performance improvement for population exceeds the number of around 100-120. The roulette wheel selection scheme [Goldberg, 1989] was chosen here, which enforces the principle of 'survival-of-the-fittest' such that individuals with higher fitness scores have higher probabilities to be selected for reproduction. The elitist strategy [Tan et al., 2001c] was used where a small portion of individuals (7 in this experiment) with good fitness scores are exempted from genetic operations and are selected directly as offspring in the next generation.

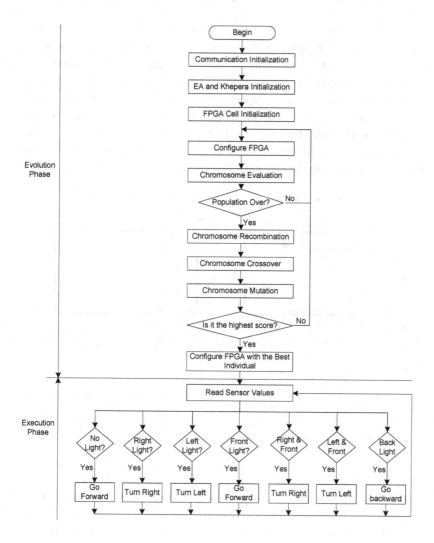

Fig. 3.12   Program flowchart of light source following task.

The elitist strategy ensures that the best individual with the highest fitness score is always survived and hence the diversity of the evolution could be preserved.

A contiguous block of cells in the FPGA is selected to perform crossover operation with a probability of 0.85 [Goldberg, 1989]. A partial rectangle with randomly selected area is chosen within the entire predefined rectangular cell array involving the evolutionary operation, and all the cell configuration bits in the pair of chromosomes are exchanged without modifying their internal genes order. Such a chromosome manipulation is in fact a multi-cut crossover operation if the cells are numbered in the order of left-to-right and top-down fashion. Fig. 3.13 illustrates the crossover operation in the hardware cells, where the two areas (4 rows and 7 columns) delimited by the solid lines are the FPGA cells configured by two different chromosomes, and the cells defined by the dashed lines denote the areas that the crossover operation takes place.

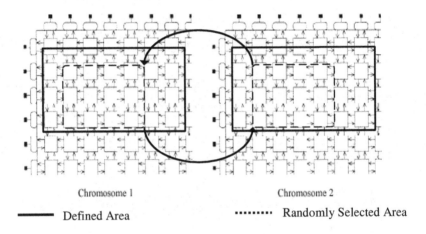

Chromosome 1   Chromosome 2

────── Defined Area   ·········· Randomly Selected Area

Fig. 3.13   Crossover operation in FPGA cell array.

From the robotic point of view, the mutation operator has three purposes, e.g., exploration, specialization and generalization. It basically introduces new robot motions for sensor input states and by this way finds the motions that avoid collisions. In this experiment, the mutation was performed at a rate of 0.01, which was chosen to be as low as the reciprocal of the chromosome length, e.g., there are 28 cells (4 rows × 7 columns) contribute to the entire evolution and each cell needs 3 bytes to configure its functionality. If the mutation rate is too high, it may introduce noise in the evaluation of the rows and cause genetic instability. If the mutation rate is too low, the robot may take a longer time to evolve the expected behaviors. The mutation rate used here is slightly higher than the inverse of the length of a chromosome, so there are in probability a few mutations performed at each generation.

The fitness of each chromosome is evaluated by treating its genes as the FPGA architecture bits, and the corresponding resources are configured to perform the intrinsic evolution. Each chromosome is presented with the same set of sensor input states and the FPGA output values are evaluated based on common sense. For instance, if any sensor on the left detects a light, the robot is supposed to make a left turn to follow the light. The output value of each infrared sensor is coded as 0 or 1 at an ambient light threshold of 250. In the implementation, the 8 sensor input states are grouped into 4 logical sensor states and the resulting 16 possible states are categorized into 7 frequently encountered situations in the physical environments. To encourage the robot to go straight when there is no light source detected, an individual scores 2 points if it responds correctly to the situation. Therefore the highest possible score is 8 points if the robot produces the desired motion in every situation, i.e., the complete light source following behavior is achieved. Fig. 3.14 illustrates how each individual is scored during the intrinsic evolution in detailed. Since there are so many possible combinations of sensory input and motion in a physical environment, a robot may only explore certain parts of the environment if the learning is carried out in a random manner. In our context, 7 different sensory inputs that represent the 7 most common world states are constantly presented to every Boolean function of the robot for behavior learning, which thus results in the shortest timeframe for evolving the task of light source following.

3.6.2.3   *Implementation of light source following task*

As shown in Fig. 3.15, the size of the physical environment for the light source following experiment is $130 \times 90 \text{cm}^2$, with four deceptive dead-ends at each corner.

The fitness convergence trace along the evolution is shown in Fig. 3.16. Each data point in the figure is an average value over 10 replications with different random initializations. As can be seen, both the average and the best fitness values converge nicely and improve along the evolution, as desired.

The FPGA hardware configuration is a 28×24 bits long binary string, e.g., 4 rows, 7 columns, and each cell contains 24 configuration bits. It represents the unknown mapping function between sensor input states and the possible motor output states. The evolution was run for 360 generations and the best fitness score at the end of the evolution is 8 points. The

Sensors (inputs)	Motors Speed (outputs)*		Scores Assigned
	Left Wheel	Right Wheel	
00000000 (0X00)	10	10	2.0
(No light detected)	Others		0.0
00010000 (0x10)	10	−10	1.0
(Right light detected)	Others		0.0
00000100 (0x04)	−10	10	1.0
(Left light detected)	Others		0.0
00001000 (0x08)	10	10	1.0
(Front light detected)	Others		0.0
00011000 (0x18)	10	−10	1.0
(Right and front light)	Others		0.0
00011100 (0x0C)	−10	10	1.0
(Left and front light)	Others		0.0
00000010 (0x02)	−10	−10	1.0
(back light detected)	Others		0.0

* Encoded speed of the Khepera robot

Fig. 3.14   Fitness evaluation in light source following task.

Fig. 3.15   Environment of the light source following experiment.

chromosome with the highest fitness score (8 points) in hexadecimal form
is shown below:

0x3C,0x8E,0x85,   0x5D,0x8D,0xCB,   0xE4,0x8C,0x85,   0x55,0x85,0xF9,
0x29,0x88,0xC4,   0x20,0x82,0x80,   0xD4,0x86,0xC8,   0xF5,0x8D,0xBD,
0xCC,0x84,0xC5,   0x6B,0x83,0xC7,   0x39,0x87,0xCE,   0x2A,0x82,0x8F,
0x8D,0x84,0x80,   0x74,0x94,0x80,   0xEA,0x84,0x07,   0x82,0x89,0xCA,
0x13,0x82,0x89,   0x24,0x89,0xC9,   0xE0,0x8B,0xCF,   0x7A,0x8E,0xC4,

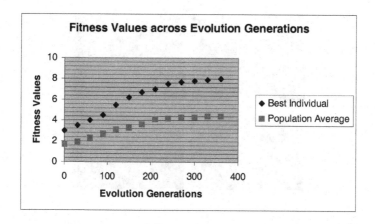

Fig. 3.16   Fitness convergence trace in light source following task.

0xD8,0x84,0xCB,   0x82,0x8D,0x8A,   0x60,0x89,0x87,   0x65,0x81,0x89,
0x49,0x80,0x81, 0x37,0x88,0xC1, 0x33,0x87,0x8A, 0x5A,0x8D,0x8D.

The intrinsic hardware evolution helps to eliminate the complexity of designing the FPGA's internal hardware structure and shift the design focus to defining the fitness function or other issues. Two physical experimental setups were performed in our study to examine the performance of the evolved robotic controller in light source following. As shown in the snapshots of Fig. 3.17, the prompt responses of the robot to the moving torch light in the first experiment demonstrate that the evolved robot is capable of following the light source located at different places in the open arena. In the second experiment, the robot is required to follow a moving light source in an S-shape as shown in Fig. 3.18. The segments in the figure represent successive displacements of the axis connecting the two wheels, and the dotted line denotes the trajectory of the light source emitted by the moving touch light. The smooth robot trajectory recorded in Fig. 3.19 shows that the EHW-based robotic controller is capable of performing light source following behavior stably and successively.

3.6.3   *Obstacle Avoidance using Robot with a Traction Fault*

In the obstacle avoidance experiment, the robot is learned to turn away from obstacles in the range of its sensors without touching it and runs straight

Fig. 3.17   Snapshots of robot motions in light source following task.

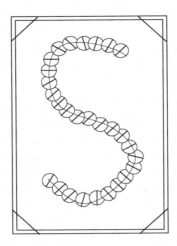

Fig. 3.18   The S-shape trajectory of light source following task.

and fast otherwise. To further evaluate the robustness and capability of the autonomous robot, one wheel of the robot is made out of control and can only keep a constant speed. Conventional robotic design methodologies cannot easily cope with the integration of fault tolerant requirements into the design process and often resort to provide spare parts. In this experiment, an external fault is introduced while keeping the robot exhibiting stable obstacle avoidance behavior during its motion. In principle, the anti-collision task is similar to the previous light source following experiment as

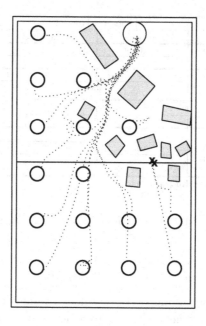

Fig. 3.19   Trajectories of the robot from 16 initial positions in the arena.

shown in Fig. 3.11. The light reflected by obstacles is detected by robot IR sensors and its value is coded into binary string to be fed into the FPGA. The outputs of the FPGA are used to guide the robot movements.

Fig. 3.20 illustrates the fitness assignment strategy for the autonomous robotic hardware evolution. The output value of each infrared sensor is coded as 0 or 1 at a proximity threshold of 500. Like the light following case, each chromosome is evaluated by treating its genes as the FPGA architecture bits, and the corresponding resources are configured to perform the desired anti-collision task. Each chromosome is presented with the same set of sensor input states and the FPGA output values are evaluated based on common sense. For instance, if any sensor on the left detects an obstacle, the robot is supposed to make a right turn to avoid collision. If the middle front proximity sensors detect anything in front, the robot can either turn right or turn left. In the experiment, a total of 6 sensor input states are presented to the FPGA. Therefore the corresponding highest fitness score is 6 points, since it scores 1 point whenever the Boolean function gives a correct response to each of the input states.

Sensors (inputs)	Motors Speed (outputs)*		Score Assigned
	Left Wheel	Right Wheel	
00000000 (0X00)	10	10	1.0
(No obstacle)	Others		0.0
00010000 (0x10)	10	20	1.0
(Right obstacle)	Others		0.0
00000100 (0x04)	10	0	1.0
(Left obstacle)	Others		0.0
00001000 (0x08)	10	0	1.0
00011000 (0x18)	10	20	
00011100 (0x0C)	Others		0.0
Front obstacle			

* Encoded speed of the Khepera robot

Fig. 3.20　Fitness evaluation in anti-collision task with an external robot fault Sensors (inputs).

3.6.3.1　Software structure of anti-collision task

All the evolutionary parameter settings here are similar to that of the light source following experiment. The program flowchart of the anti-collision experiment using a robot with a traction fault is illustrated in Fig. 3.21. One merit of the proposed evolvable robotic controller design approach is that the evolution and execution are separated into two successive phases. The evolved architecture bits for a particular robotic task can be regarded as a general purpose mapping between the FPGA inputs and outputs, and hence can be utilized by another robotic task without much revision, e.g., only the execution phase of the program is needed to be revised. For example, similar to the program of light source following, the optimal architecture bits evolved in the learning phase is used to configure the FPGA in the execution phase to perform the desired task. Prior to the robotic behavior execution, a fault is simulated in the physical Khepera robot by setting its left wheel speed to a constant of 10 units, and the robot is only permitted to adjust the right wheel speed for obstacle avoidance in the experiment.

3.6.3.2　Implementation of obstacle avoidance task

As shown in Fig. 3.22, the EHW-based robotic controller is validated upon obstacle avoidance experiments in a physical environment. As can be seen, the colors of the walls and obstacles are chosen in white in order to give a perfect reflection to the infrared light sensor. The obstacle shapes are designed in cubic so as to avoid the robot being stuck during its motions

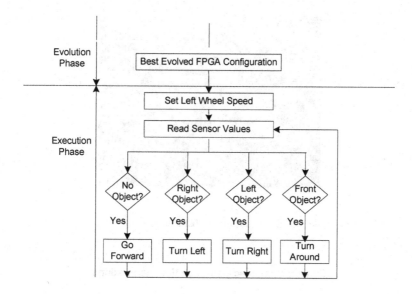

Fig. 3.21   Program flowchart of anti-collision task with an external robot fault.

in the experiment. Fig. 3.23 depicts the fitness convergence trace for the anti-collision experiment. Each data point in the figure is an average value over 10 replications with different random initializations. The maximum fitness score obtained at the end of the evolution is 6 points, as desired. Similar to the evolution for light source following, it can be seen from Fig. 3.23 that both the average and the best fitness values converge nicely and improve along the evolution.

The hexadecimal data given below represent the best evolved architecture bits (with the fitness score of 6), which can be stored and manually configured in the FPGA:

0xB2,0x8E,0xAA,   0x8C,0x85,0x83,   0x5E,0x8F,0x83,   0x19,0x8B,0x89,
0xFF,0x88,0x8E,   0xC3,0x8E,0xCB,   0x99,0x80,0xC8,   0xDE,0x82,0x83,
0x4F,0x85,0xC7,   0xE2,0x8A,0x82,   0x7E,0x85,0xC4,   0xBD,0x8C,0xC9,
0x5F,0x87,0x8F,   0x65,0x80,0xE6,   0x43,0x00,0x87,   0x4F,0x82,0x88,
0x0E,0x85,0xC0,   0x5C,0x80,0xC6,   0xA6,0x80,0xCA,   0xB9,0x82,0x85,
0xCA,0x8A,0x8A,   0xA7,0x81,0xCE,   0x46,0x8A,0xCA,   0x02,0x87,0x80,
0x17,0x87,0xCF, 0x38,0x86,0x8D, 0x7E,0x88,0x80, 0x12,0x8F,0xC6.

Two physical experiments were carried out to examine the evolved robotic controller, e.g., the tasks of robot corridor following and robot

Fig. 3.22   Environment of the robot obstacle avoidance implementation.

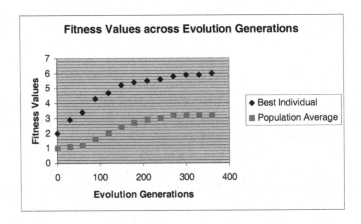

Fig. 3.23   Fitness convergence trace in anti-collision task.

movements in an arena with random distributed obstacles. In the first experiment, the corridor following task was conducted to illustrate the robotic behavior of performing straight navigation while avoiding any obstacles encountered in the environments. The robot was put in the environment consists of a circular corridor as shown in Fig. 3.24. The walls are made of light yellow woods and the floor is black. In Fig. 3.25, segments represent successive displacements of the axis connecting the two wheels, and the direction of the motion is clockwise. As can be seen, the robot is able to perform complete laps along the corridor without turning back or bumping into the walls or corners.

The second experiment was carried out in an open arena with many distributed white obstacles. The complete trajectory produced by the robot is shown in Fig. 3.26. Clearly, the anti-collision behavior of the robot is robust and capable of coping with various shapes of obstacles randomly situated in the physical environment. The robot succeeds in maintaining straight movements whenever possible while keeping the navigation trajectory extremely smooth without touching any obstacles, which are primarily contributed by the novel fitness assignment scheme proposed in this chapter.

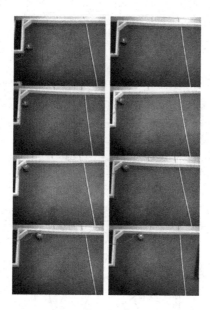

Fig. 3.24   Robot motions in avoiding the arena corner.

3.7   Summary

This chapter has presented the design and real-time implementation of an intrinsic evolvable hardware-based autonomous robot navigation system. The concept of Boolean function has been used to construct the evolvable controller implemented on an FPGA-based robot turret, and evolutionary computing has been applied as a learning tool to guide the artificial evolution at the hardware level. It has turned out that a mapping of a human-

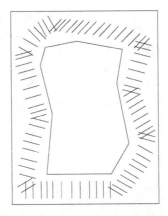

Fig. 3.25   The robot trajectory in corridor following task.

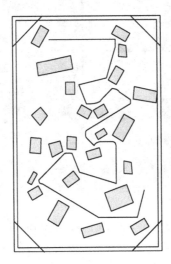

Fig. 3.26   The robot trajectory in obstacle avoidance task.

specified Boolean function onto the FPGA hardware could be evolved successfully. The effectiveness of the proposed evolvable autonomous system
have also been demonstrated via physical implementation of robot navigation behaviors on light source following and obstacle avoidance using a
robot with traction fault.

This chapter has been a manifesto for an approach to evolving hardware for robot control. As promised at the introduction, the conclusion is that points one to three of my thesis have been shown to be true, using a combination of theory, verbal argument, and empirical demonstrations:

- The concept of an evolvable Boolean function controller at the hardware level using FPGA is demonstrated to be feasible. In the experiment, we have shown and demonstrated that the Boolean controller using logic-function based EHW was able to build off-line a highly robust control system which is not very seriously affected by the shape of the obstacles and the position of the light source.

- Intrinsic hardware evolution can be allowed to explore hardware resources more efficiently than the conventional design. With their richer dynamic behavior and less constrained spatial structure, the evolved circuit may perform better in a different way that electronics is normally envisaged. The increased freedom allows evolution to exploit the properties of the hardware resources more effectively in achieving the desirable task. Taking an FPGA intended to be used in a digital way, and omitting some constraints needed to support the digital design methodology, a satisfactory behavior has been evolved in an extraordinarily small silicon area.

- Offline evolution can build the robust robotic behaviors in real robot implementation. The successful transfer of robotic behaviors from evolution phase to execution phase strongly demonstrated that the robotic behaviors obtained by offline evolution can execute in the physical environment stably and elegantly.

The research has aimed to explore the new field of intrinsic hardware evolution in robotic controller design, by investigating the relationships with existing knowledge: conventional electronics design, and natural evolution. The former is a different process in the same medium, and the latter is a similar process in a different medium. They are both rich sources of techniques and inspirations, but intrinsic hardware evolution is a new combination of process and medium, and its full potential can only be realized by exploring the new forms that are natural to it.

From this study, we can see that the concept of evolvable Boolean function controller at the hardware gate-level using FPGA is feasible, which offers an effective autonomous robotic system that is not seriously affected by the shapes of the obstacles, the positions of the light source, and the external failure of the robot. In addition, the approach of off-line hardware evolution is robust in physical autonomous robotic system. The successful transfer of robotic behaviors from the evolution phase to the execution phase has demonstrated that navigation behaviors obtained by the off-line evolution can execute in the physical environment stably and elegantly.

Whether or not the potentially great engineering benefits of fully unconstrained intrinsic hardware evolution turn out to be completely realizable in practical applications, the fundamental groundwork developed herein must provide some meaningful information for the future development of the field.

Bibliography

de Garis, H. (1996). CAM-BRAIN: The evolutionary engineering of a billion neuron artificial brain by 2001 which grows/evolves at electronic speeds inside a Cellular Automation Machine (CAM). *In Sanchez, E., and Tomassini, M. (Eds.), Towards Evolvable Hardware: The evolutionary engineering approach, vol. 1062 of LNCS*, pp. 76–98. Springer-Verlag.

Fogel, D. B. (2000). What is evolutionary computation? *IEEE Spectrum*, pp. 26–32.

Franzi, E. (1998). *Khepera BIOS 5.0 Reference Manual*, K-Team, S. A.

Goldberg, D. E. (1989). *Genetic algorithms in search, optimization and machine learning*, Addison Wesley.

Grefenstette, J. J. (1996). Genetic learning for adaptation in autonomous robots, *Robotics and Manufacturing: Recent Trends in Research and Applications.* ASME Press, vol. 6, New York.

Haddow P., and Tufte, G. (1999). Evolving a robot controller in hardware, *Norwegian Computer Science Conference (NIK'99)*, pp. 141–150.

Higuchi, T., Iwata, M., Kajitani, I., et al. (1997). Evolvable hardware and its applications to pattern recognition and fault-tolerant systems, *Towards Evolvable Hardware: the evolutionary engineering approach, Lecture Notes in Computer Science 1062*, pp. 118–135, Springer-Verlag.

Higuchi, T., Kajihara, N. (1999). Evolvable hardware chips for industrial applications, *Communications of the ACM*, vol. 42, no. 4, pp. 60–69.

Higuchi, T., Iwata, M., Keymeulen, D., et al. (1999). Real-world applications of analog and digital evolvable hardware, *IEEE Transactions on Evolutionary Computation*, vol. 3, no. 3, pp. 220–235.

Husbands, P. (1998). Evolving robot behaviors with diffusing gas networks, *Proceedings of the First European Workshop on Evolutionary Robotics 98 (EvoRobot98)*, France, pp. 71–86.

Iwata, M., Kajitani, I., Yamada, H., et al. (1996). A pattern recognition system using evolvable hardware, *Parallel problem solving from nature - PPSN IV, Lecture Notes in Computer Science 1141*, pp. 761–770, Springer-Verlag.

Jakobi, N. (1997). Half-baked, ad-hoc, and noisy: minimal simulations for evolutionary robotics, *In Husbands, P., and Harvey, I., (Eds.) Proceeding of*

Forth European Conference on Artificial Life, MIT Press.

Jakobi, N. (1998). Running across the reality gap: Octopod locomotion evolved in a minimal simulation, *Proceedings of the First European Workshop on Evolutionary Robotics 98 (EvoRobot98)*, France, pp. 39–58.

Jakobi, N., and Quinn, M. (1998). Some problems (and a few solutions) for open-ended evolutionary robotics, *Proceedings of the First European Workshop on Evolutionary Robotics 98 (EvoRobot98)*, France, pp. 108–122.

Keramas, J. G. (2000). How will a robot change your life? *IEEE Robotics and Automation Magazine*, pp. 57–62.

Keymeulen, D., Konaka, K., Iwata, M., et al. (1997). Robot learning using gate-level evolvable hardware, *Proceeding of Sixth European Workshop on Learning Robots (EWLR-6)*, Springer Verlag.

Kitano, H. (1996). Challenges of evolvable systems: analysis and future directions, *Evolvable Systems: From Biology to Hardware, Lecture Notes in Computer Science 1259 (Proc. of ICES1996)*, pp. 125–135, Springer-Verlag.

Manderick, B. and Higuchi, T. (1996). Evolvable hardware: an outlook, *Evolvable Systems: From Biology to Hardware, Lecture Notes in Computer Science 1259 (Proc. of ICES1996)*, pp. 305–310,

Springer-Verlag. Mondada, F., Franzi, E., Ienne, P. (1993). Mobile robot miniaturization, *Proceedings of the Third International Symposium on Experimental Robotics*, Kyoto, Japan.

Nolfi, S., and Floreano, D. (2000). *Evolutionary robotics: biology, intelligence, and technology of self-organizing machines*, Cambridge, MA: MIT Press.

Porter, R., McCabe, Bergmann, N. (1999). An application approach to evolvable hardware, *Proceedings of the First NASA/DoD Workshop on Evolvable Hardware*, California, IEEE Computer Society Press, pp. 170–174.

Sanchez, E., Mange, D., Sipper, M, et al. (1996), Phylogeny, ontogeny, and epigenesis: three sources of biological inspiration for softening hardware, *Evolvable Systems: From Biology to Hardware, Lecture Notes in Computer Science 1259 (Proc. of ICES1996)*, pp. 35–54, Springer-Verlag.

Seth, A. K. (1998). Noise and the pursuit of complexity: A study in evolutionary robotics, *Proceedings of the First European Workshop on Evolutionary Robotics 98 (EvoRobot98)*, France, pp. 123–136.

Shirasuchi, S. (1996). FPGA as a key component for reconfigurable system, *Evolvable Systems: From Biology to Hardware, Lecture Notes in Computer Science 1259 (Proc. of ICES1996)*, pp. 23–32, Springer-Verlag.

Smith, T. M. C. (1998). Blurred vision: Simulation-reality transfer of a visually guided robot, *Proceedings of the First European Workshop on Evolutionary Robotics 98 (EvoRobot98)*, France, pp. 152–164.

Tan, K. C., Lee, T. H., Khoo, D. and Khor, E. F. (2001a). A multi-objective evolutionary algorithm toolbox for computer-aided multi-objective optimization, *IEEE Transactions on Systems, Man and Cybernetics: Part B (Cybernetics)*, vol. 31, no. 4, pp. 537–556.

Tan, K. C., Lee, T. H. and Khor, E. F. (2001b). Automatic design of multivariable QFT control system via evolutionary computation, *Proc. I. Mech. E., Part I*, vol. 215, pp. 245–259, 2001.

Tan, K. C., Lee, T. H. and Khor, E. F. (2001c). Evolutionary algorithm with dynamic population size and local exploration for multiobjective optimization, *IEEE Transactions on Evolutionary Computation*, vol. 5, no. 6, pp. 565–588, 2001.

Thompson, A. (1995). Evolving electronic robot controllers that exploit hardware resources. *Proceeding of the 3rd European Conf. on Artificial Life (ECAL95)*, pp. 640–656, Springer-Verlag.

Thompson, A., Layzell, P., and Zebulum, R. S. (1999). Explorations in design space: unconventional electronics design through artificial evolution, *IEEE Transactions on Evolutionary Computation*, vol. 3, no. 3, pp. 167–196.

Wang, L. F., Tan, K. C., and Prahlad, V. (2000). Developing Khepera robot applications in a Webots environment, *Proceedings of the 11th IEEE International Symposium on Micromechatronics and Human Science*, Japan, pp. 71–76.

Xilinx Inc. (1996). *XC6200 Product Specification V1.0*, 1996 (Http://www.xilinx.com).

Yao, X. and Higuchi, T. (1999). Promises and challenges of evolvable hardware, *IEEE Trans. on Systems, Man, and Cybernetics- Part C: Applications and Reviews*, vol. 29, no. 1, pp. 87–97.

Chapter 4

Intelligent Sensor Fusion and Learning for Autonomous Robot Navigation*

This chapter presents the design and implementation of an autonomous robot navigation system for intelligent target collection in dynamic environments. A feature-based multi-stage fuzzy logic (MSFL) sensor fusion system is developed for target recognition, which is capable of mapping noisy sensor inputs into reliable decisions. The robot exploration and path planning are based on a grid map oriented reinforcement path learning system (GMRPL), which allows for long-term predictions and path adaptation via dynamic interactions with physical environments. In our implementation, the MSFL and GMRPL are integrated into subsumption architecture for intelligent target-collecting applications. The subsumption architecture is a layered reactive agent structure that enables the robot to implement higher-layer functions including path learning and target recognition regardless of lower-layer functions such as obstacle detection and avoidance. The real-world application using a Khepera robot shows the robustness and flexibility of the developed system in dealing with robotic behaviors such as target collecting in the ever-changing physical environment.

4.1 Introduction

Target collection using autonomous robots can be broadly applied to a variety of fields such as product transferring in manufacturing factory, rubbish cleaning in office, and bomb searching on battle field, etc. Such robots should be able to cope with the large amount of uncertainties exist in physical environment which is often dynamic and unpredictable. The task

---

*Copyright (2005) From ("Intelligent sensor fusion and learning for autonomous robot navigation") by (Tan, K. C., Chen, Y. J., Wang, L. F. and Liu, D. K.). Reproduced by Taylor & Francis Group, LLC., http://www.taylorandfrancis.com

requires the robot to achieve certain goals without bumping into obstacles based on sensor readings. However, the goal-based world model is often difficult to be precisely predefined and sometimes the obstacles are not static. Moreover, the sensor information is imprecise and unreliable in most cases. Therefore, the autonomous robot controller needs to be endowed with the capability of building world model by itself and making reliable decisions under various uncertainties in order to operate properly in the real-world environment.

A commonly used strategy of designing an adaptive and flexible controller for certain tasks is to decompose the complex task into small independent decision-making agents and each agent fully implements a control policy for one specific sub-task. These agents gather data through their sensors and react to certain events [28]. Traditionally, the agents are classified as projective agents and reactive agents [27]. In our autonomous robot implementation, the projective agents and reactive agents are designed as knowledge layer and reactive layer respectively in a hybrid architecture [21], which is a mixture of the deliberate architecture and reactive architecture [22]. This strategy endows the robot the capability of reasoning about future states over long term as well as responding to real-time emergencies. The projective agents are goal-oriented requiring underlying model. It develops plans and makes decisions for a long-term purpose. Conversely, the reactive agents are sensor-based without complex reasoning and it is responsible for real-time performance. Moreover, it may override the projective agents if necessary [28].

In the layered hybrid architecture, the higher-level abstract layers are built and relied upon lower-level concrete layers. The lower-level functions typically have higher priorities over the higher-level ones, i.e., the lower level behavior such as obstacle detection should override higher level functions such as path following so that the robot would not collide into the wall or obstacles. To design the controller with the multi-agent architecture in mind, the various layers of competence are identified and each layer is built from the bottoms up, incrementally. Newer designs are added to the existing and functioning designs. The higher-level modules have access to lower-level data and can send data to lower levels to suppress normal data flow. The interaction between layers is kept minimal to achieve maximum "incrementality" and emergence, though links still have to be added between layers to inhibit certain behaviors. Each agent is a finite state machine (FSM) [20], that is, the next state is associated with the current state and input. The architecture enables the implementation of higher layer functions of target recognition and path planning incrementally on top of lower layer functions of basic robot maneuvering and obstacle avoidance. In our study, we put emphasis on the design of knowledge layer projective agents, which is the main challenge on today's autonomous robotics. More strategies are needed to deal with uncertainties in the physical environment so that reliable decisions can be made in real-time.

In the robotic task of target collection,target recognition and path planning are the two challenging projective agents to be implemented. Target recognition under uncertainties is difficult to realize since the sensor information is too noisy to be directly mapped to a definitive representation of the target. It has been widely studied in recent decades and multi-sensor data fusion [24] has proved to be an effective approach to tackle this problem. Sensor data fusion is the combination and integration of inputs from different sensors, information processing blocks, databases or knowledge bases, into one representational format [1]. It yields more meaningful information as compared to the data obtained from any individual sensor module [2]-[3]. Most fusion systems in robotics are intended for applications such as pattern recognition [4]-[5] and localization [6]-[7]. The proposed techniques for sensor data processing include the least square method, Bayesian method, fuzzy logic, neural networks and etc. Each of these artificial intelligence methods has their own merits [11], [23]. For instance, one of the promising fusion techniques named fuzzy logic has the advantage of mapping imprecise or noisy information as input into an output that enables precise decision to be inferred from it [13], [25]-[26]. It is especially useful when the data is obtained under uncertain conditions. Their operation and inference rules are relatively clearer, which can be described in natural languages in the form of linguistic variables [8]-[9]. In our work, a multi-stage fuzzy logic (MSFL) inference system is deployed to accelerate the recognizing process. In this system when an object is identified as a non-target in the lower stage, no actions will be taken in the higher stages which need further sensor information.

Another projective agent for path planning makes use of reinforcement learning mechanism [12] to find the shortest route to the designated destination via self-learning rather than supervised learning. Therefore, the robot is able to plan path through sensor information and self-learning algorithm instead of a predefined model which is normally non-flexible and non-adaptive. Reinforcement learning focuses on its goal more than other approaches to machine learning [19]. It is to map situations to actions. At each time step, the learning system takes an action after receiving the updated information from the environment. It then receives a reward in the form of a scalar evaluation. The system selectively retains the outputs that maximize the received reward over time [12]. The four key elements of reinforcement learning are the policy, reward, value, and model. According to Sutton [12], policy is the agent's way of behaving at a given time. Reward is the intrinsic desirability of the state. It has to maximize reward in the long run. Value of a state is the total amount of reward accumulated over the future from that state onwards. It is a measure of what is good in the long-term run. The last element is the model of the environment. In real world, the environments are characterized by uncertainties of unpredictable events. Furthermore, a prior defined model cannot adapt to a changing environment easily. Therefore, the robot should be able to build and update this world model autonomously according to the real-time information from sensors. The central idea of reinforcement learning is called temporal difference (TD) learning [14], which can make long-term predictions about dynamic systems. In learning process, it estimates how good it is to take a certain action in a given state through trial-and-error search and delayed reward. The essence of reinforcement learning lies in its self-adaptive capability by interacting with its environment to achieve the goal. The issue of exploration and exploitation arises here is also balanced. In other words, an agent has to exploit or maximize what it already knows yet it has to explore for better actions to make in the future.

The remaining part of the chapter is organized as follows: Section 4.2 introduces the experiment platform including Khepera robot [16] and Webots simulation software [17]. The hybrid architecture of the controller is also described in Section 4.2. Section 4.3 presents details of the MSFL inference system for target recognition agent. The GMRPL path planner for path planning agent is discussed in Section 4.4. Experimental results are given in Section 4.5 and conclusions are drawn in Section 4.6.

4.2   Development Platform and Controller Architecture

4.2.1   *The Khepera Robot and Webots Software*

As shown in Fig. 4.1, the Khepera robot constructed in modular concept [16] is selected as the hardware platform so that additional function modules can be plugged into the base robot easily. The Kepera is equipped with eight IR sensors, six in the front and two in the back, which return the proximity of objects around the robot. The distribution of the IR sensors is shown in Fig. 4.2. Khepera has two wheels, which are individually driven by DC motors. The speed of each wheel is controlled by a PID controller and on each wheel an incremental encoder is placed to calculate the robot's position. In addition, the K6300 2-D camera module [16] for acquiring image and the gripper module [16] for grasping the object are mounted on the base platform. To simplify the design and test of the robotic controller, the Webots 2.0 software [17], a 3-D GUI-based simulator is deployed.

Fig. 4.1   The Khepera robot with K6300 Turret.

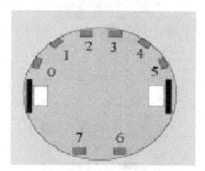

Fig. 4.2   Layout of the 8 IR sensors on Khepera.

In the software, the virtual environment can be easily constructed and managed. The sensor information can be directly viewed through intuitive user interface and the communication between different modules can be effectively monitored using a supervisor controller [17] based on the External Authoring Interface (EAI) in Webots. An example GUI is shown in Fig.

4.3. The IR sensors' value and image information from the camera of the robot is shown in Fig. 4.3(a) and Fig. 4.3(b), respectively. It can be seen that the robot is facing a can in the environment as shown in Fig. 4.3(c). The front two IR sensor values show the proximity between the robot and the obstacle, which are larger than that of others in Fig. 4.3(a). This reveals the corresponding position of the obstacle between them. In Fig. 4.3(b), the image shows the color and shape of the obstacle.

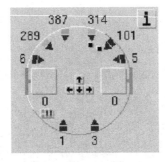

Fig. 4.3 (a) IR sensor readings; (b) Image information; (c) A sample Webots simulation GUI.

4.2.2 *Hybrid Architecture of the Controller*

To implement the robotic behavior of target collecting in a dynamic environment, the robot seeks to construct a map of its environment, locate itself in that map, plan path to the goal, and recognize objects to be collected. Self-location, path planning, target recognition, motion control, and vehicle dynamics should be coordinated by the robotic controller, which is divided into the following different modules:

(1) moving in the environment without colliding with the obstacles;
(2) setting up the grid map and locating itself in the world via position computing;
(3) extracting useful data from the raw sensor readings;
(4) integrating the sensor readings for target recognition;
(5) planning the optimal path to destination.

These modules are implemented as agents in corresponding layers in a hybrid architecture as shown in Fig. 4.4. In this architecture, the higher layers can subsume or inhibit the lower layers. The lowest level is the reactive layer consisting of reactive agents. It makes decisions about what to do according to raw data from the sensors. This layer is typically implemented similar to the Brooks subsumption architecture [10], one of the first reactive architecture models. The middle level is called knowledge layer comprised of projective agents. It adopts symbolic representations extracted from raw data like deliberate architectures in which decisions are made via logical reasoning. Above these two layers, the highest level is a social knowledge layer in which an arbitration strategy is formed to decide which agents should be activated depending on the current task goal and the environmental contingencies.

The agents functionality in each layer is illustrated in Fig. 4.5. At layer 1, the basic behaviors of obstacle avoidance and path planning enable the robot to navigate in the environment safely. At the layer 2, the robot is given the ability to locate itself in the environment so that it can determine its current state. This ability is realized using position computation based on the encoder's data on the robot wheel and the world grid map being set up with a global camera. In this layer, no consideration concerning obstacle avoidance is needed since it is the responsibility of layer 1. At layer 3, the object's physical attributes are extracted from sensor readings, which are the inputs for the target recognition system at the layer 4. By communicating with and controlling the sensor modules of camera and

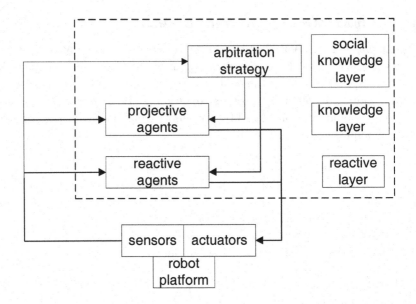

Fig. 4.4   The hybrid architecture for autonomous robot implementation.

gripper, the object is detected, located and the camera image is processed to obtain useful data, which is combined with sensors information obtained from the gripper. With all functions provided by these lower layers, at layer 5 the robot plans its optimal path via reinforcement learning [12] and moves forward without colliding with obstacles, and locates itself rightly in the grid map of the world.

As depicted in Fig. 4.6, the data flow can be described as follows:

- Step 1: obtain the grid map of the world via the global camera. The image information is represented as a matrix of $G[I][J]$ based on pixels, which is then transformed into a normalized virtual grid map $M[X][Y]$ representing the occupancy situation in each physical grid;
- Step 2: proximity information $P(i_1, i_2, \ldots, i_8)$ about the objects nearby is retrieved and the vehicle motor speed (V_L, V_R) is set according to the safe moving guideline;
- Step 3: during the navigation in the real world, the robot computes its current location $(X(t), Y(t))$ based on its previous location $(X(t-1), Y(t-1))$. The distance value $D(t)$ is retrieved from encoders on the

Layer 5: Optimal path plan	Reinforcement path planning				
Layer 4: Target recognition	Multi-stage fuzzy logic sensor fusion				
Layer 3: retrieve features	Camera control			Gripper control	
	Get colour	Get height	Get light	Get size	Get resistivity
Layer 2: Self-location	Grid map set up			Position computation	
Layer 1: Safe moving	Obstacle avoidance			Path following	
	Go forward	Go backward		Turn left	Turn right

Fig. 4.5   The hybrid architecture layer design.

wheel and its steering-wheel angle (t):

$$X(t) = X(t-1) + D(t)cos((t)) \tag{4.1}$$

$$Y(t) = Y(t-1) + D(t)sin((t)). \tag{4.2}$$

- Step 4: physical features $\{x_1 = \text{light}, x_2 = \text{color}, x_3 = \text{height}, x_4 = \text{resistivity}, x_5 = \text{size}\}$ of the detected objects are retrieved from video camera's image represented as $C[I][J]$ and the values from resistivity sensor and position sensor on the gripper;
- Step 5: the physical feature vector $\{x_1, x_2, \ldots, x_5\}$ is fused into a target recognition decision, based on which the robot chooses its corresponding behavior. This step is implemented by an MSFL inference system which will be described in Section 4.3;
- Step 6: when the robot decides to go to the appointed destination, the optimal path is computed based on the grid world map $M[X][Y]$. The path is determined by $A(x(t), y(t))$, which represents the goal-oriented optimal motion control $A = \{\text{back, forward, left, right}\}$ at location $(x(t), y(t))$. This step is implemented by a GMRL path planner which will be described in Section 4.4.

4.2.3   Function Modules on Khepera Robot

In processing the sensor information, the information on IR sensors and wheel encoders is directly retrieved from the base Khepera. The video

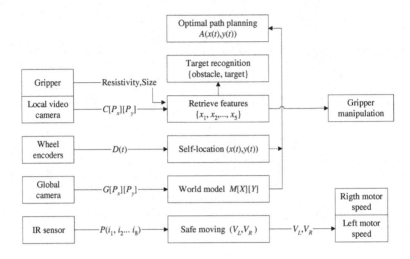

Fig. 4.6   Data flowchart of the working system.

camera and gripper are adds-on modules mounted on the base robot and both communicate with the base through a virtual supervisor module implemented by Webot's EAI. The data from these two modules is preprocessed before transferred to the projective agents. The detailed flow diagram describing the communication between different modules is shown in Fig. 4.7. Therefore, there are three function modules mounted on the base Khepera robot, e.g., two physical turrets of K6300 camera and gripper, and a virtual supervisor module controlling their communications. When an obstacle is detected, the K6300 adjusts its direction to face the object and checks if it is the targeted object (e.g., a can). If the can is detected by its two front sensors, it orientates itself to a standard position to read light intensity, height, and the RGB color components before sending the data to the supervisor. When the supervisor receives the data from the K6300 module, the supervisor uses these values as inputs to the target recognition system. If the object is recognized as a non-target, the supervisor labels it as an obstacle in the grid map and commands the robot to seek new objects. If the object matches the target's visual representation, the gripper module is interchanged with the K6300. When detecting any objects, the robot reads resistivity and size of the object from the gripper arm sensor and such data is sent back to the supervisor. The gripper waits for an answer from the supervisor's sensor fusion module. When the target recognition system determines that the target is detected, the gripper module grips the can and brings it to the goal position using the shortest path gained through reinforcement path learning.

The supervisor in the controller acts as the highest layer in the hybrid architecture. It decides the agent to be activated and manages the communications between different function modules. That is, the module not only controls the behaviors of the base Khepera, the K6300, and gripper turrets, but also transfers information to the fuzzy logic target recognition system and reinforcement path learning system. It determines the robot's position in the environment and manages the grid map to adapt to any changes in the environment. The stage of the fuzzy logic recognition process is also selected by the supervisor controller. When any controller is activated by the supervisor, it re-starts again.

4.3 Multi-Stage Fuzzy Logic (MSFL) Sensor Fusion System

4.3.1 *Feature-Based Object Recognition*

In this work, sensor data is selected according to different features characterizing the physical attributes of the detected object. These inputs are fused into previously formed representations of the real world target. Thus, the target recognition problem can be summarized as follows:

- The attributes set of the observed (unknown) object;
- The definition of the target;
- The fuzzy logic inference system mapping the sensor inputs into the definition, embracing fuzzy sets for each attribute, fuzzy logic operator, and the 'if-then' rule base.

Assuming a target can be identified on the basis of 5 different features $\{x_1, x_2, \ldots, x_5\}$ which are obtained from the sensor modules on the robot, and each feature $x_j (j = 1, 2, \ldots, 5)$ has k number of fuzzy predicates $\{a_{j1}, a_{j2}, \ldots, a_{jk}\}$. If the target is represented as $\{c_{i1}, c_{i2}, \ldots, c_{i5}\}$, where $i = 1, 2, \ldots, 5$, N is the number of defined targets, and $c_{ij} \in \{a_{j1}, a_{j2}, \ldots, a_{jk}\}$, the fuzzy set rules which yield the degree of membership of the detected object to the predefined target could be formulated as follows,

R_i: *IF* $(x_1$ *is* $C_{i1})$ *AND* $(x_2$ *is* $C_{i2})$ ... *and* $(x_5$ *is* $C_{ir})$ *THEN observed object is* C_i

where C_i is fuzzy sets for the observed object. Based on the Mamdani inference model and that "a AND b" = MIN(a, b) as well as "a OR b" =

MAX(a, b), the given inputs are in the form of

x_j is A'_j

where A'_j are fuzzy subsets of $U_1 \times U_2 \times \ldots \times U_r$, the contribution of rule R_i to a Mamdani model's output is a fuzzy set whose membership function is computed by,

$$\mu_{c'_i} = (\alpha_{i1} \cap \alpha_{i2} \cap \ldots \cap \alpha_{ir}) \cap \mu_{c_i}(y) \qquad (4.3)$$

where α_i is the matching degree (i.e. firing strength) of rule R_i, and where α_{ij} is the matching degree between x_j and R_i's condition about x_j:

$$\alpha_{ij} = \sup_{x_j} \{\mu_{A'_j}(x_j) \cap \mu_{A_{ij}}(x_j)\} \qquad (4.4)$$

and $\cap$ denotes the "min" operator. The final output of the model is the aggregation outputs from all rules using the max operator:

$$\mu_c(y) = \max\{\mu_{c'_1(y)}, \mu_{c'_2(y)}, \ldots, \mu_{c'_L(y)}\}. \qquad (4.5)$$

The output C is a fuzzy set which is defuzzified to give a crisp output.

4.3.2 MSFL Inference System

In order to accelerate the target recognition, an MFSL inference system is adopted as shown in Fig. 4.8. At the lower stage, if an observed object can be identified as a non-target based on the lower stage sensor inputs, it will not go on to the higher stage fuzzy sensor fusion. The whole inference system is divided into three stage sub-systems. The output from one stage in fuzzy inference system is taken as an input to the next stage. At the stage 1, the object color is determined based on the RGB (red, green, blue) components of the object. If it does not match the target's color definition, the system views it as a non-target, and asks the supervisor to seek for another object. If the color matches, the system retrieves the light and height information of the object from the supervisor and combines them with the color decision from stage 1 to determine the visual likeness between the object and the target. If all the three attributes (i.e., color, light, and height) match the target representation, the resistivity and size of the object are added together with the visual decision as inputs to the stage 3 to obtain the final decision on target recognition. The whole system works step by step, e.g., the lower stage decision is the base for higher stages. When more sensors are added to the robot to collect other physical parameters, the system can scale with new stages added to the current system.

As shown in Fig. 4.9, the shape of membership function used in the fuzzy system depends on the range of the sensor input values as given in Fig. 4.10. When different sensors are deployed, the membership function needs to be modified accordingly.

In the first stage, the fuzzy inference system checks if the color of pre-assigned target matches that of the object. As shown in Fig. 4.11, the object color is red in this experiment. The input membership functions of each RGB color components are two triangular membership functions with the values of "small" and "large" according to the intensity. The RGB information extracted from camera sensor is first normalized into variable domain, e.g., [0, 255]. This crisp value is then fuzzified to determine the degree of which they belong to the appropriate fuzzy sets via the membership functions of small and large. Adopting fuzzy approximate reasoning to infer the color recognition decision contributed from each rule and aggregating the fuzzy outputs of the red, blue and green degree, the output color fuzzy set is obtained which is then defuzzified and denormalized into the required domains. The final output is passed to a criterion that determines the actions of the robot. Using the center of area method, the set is defuzzified to a real value varies from 0 to 80 depending on the contribution of the color component membership function. This fuzzy logic method can be used as an alternative to filtering method in color recognition. By observations, the value over 75 is considered as redness and is hereby used in this stage. The degree of color mixture to come to the amount of redness depends on the setting of the membership function at the input. Thus, the perception of redness and the value vary with different sources and subjects. The recognition for other colors can be easily applied using this method by rotating the color ramp and rearranging the output membership functions.

The initial membership functions are constructed arbitrarily with triangular and trapezoidal shapes. Depending on the application and the specification of the designated target, the parameters of the membership functions at any stage may vary. At the output membership function, the overlapping is constructed proportional to the mixture of primary colors (red, green, and blue) that forms the color ramp. Three rules are used in the color identification, which are:

IF red is large AND green is small AND blue is small, THEN target color is red IF red is small AND green is large AND blue is small, THEN target color is green IF red is small AND green is small AND blue is large, THEN target color is blue.

At the second stage, the sensor data concerning the light and height of the observed objects are gathered from the K6300 camera and fused with the output color value from the first stage. Target color from the first stage, raw height, and light intensity are fed to the second stage. Two output functions are constructed. The rules that are being fired are:

- *If light is large AND height is large AND targetcolor is red, THEN object is look like target.*
- *If light is small OR height is small OR targetcolor is blue OR targetcolor is green, THEN object is not look like target.*

The use of fuzzy connective 'OR' greatly reduces the computation demands as compared with the use of 'AND' since in our case the target is defined as having large values. If any value is small, it might not be the target. However, if any intermediate functions and values are used to recognize the target, more rules should be added. The final crisp value is obtained using the center of area defuzzification method. The objects are classified into target or non-target by a preset threshold.

At the third stage, the sensor data concerning the resistivity and size of the detected object are retrieved from the gripper through contact and fused with the output value from the second stage. The visual recognition value from the second stage, the raw value of resistivity and size are input into the third stage. Two output functions are constructed. The rules that are being fired are:

- *If resistivity is large AND size is large AND visual recognition is high, THEN object is target.*
- *If resistivity is small OR size is small OR visual recognition is low, THEN object is not target.*

When more parameters are needed to define the target, some sensors should be added to the robot navigation system. However, the former fuzzy inference system still works as lower stages in the new application without any modifications.

4.4 Grid Map Oriented Reinforcement Path Learning (GMRPL)

4.4.1 *The World Model*

A grid map world model [18] describing the experimental arena is constructed. The physical environment is mapped onto a regular grid, which stores value of occupancy possibility of objects in the world. In this work, the world environment that the agent interacts on is a 0.8 meter by 0.8 meter square area and rimmed by a 0.1 meter wall along its perimeter. The environment is constructed simulating a real office room with obstacles scattering around the area. The reinforcement learning utilizes the generated grid map of this world to obtain the shortest path. At first, the world grid map is captured by a webcam situated on top of the world. During its motion, the robot locates itself in the environment and modifies the grid map value when detecting any new obstacle using this model. The captured world image is processed to obtain a world grid map with value in each grid. It is first normalized and divided into 10 by 10 grid unit inclusive of the 0.1 meter wall. Thus, the 10×10 square is made up of 8×8 units of space and a width of one grid unit at each edge as the border. Each grid unit measures 10 cm by 10 cm on the real world. In the experiment world, different class of objects are identified with different colors, such as blue-colored obstacles, black blank areas, green start and red goal positions are identified by the normalizing program.

In the normalization procedure, the program crops out the target region from the picture after identifying the corner making points. Then all RGB pixel values for all bytes on each grid is summed up and taken average of it. According to the preset thresholding rule, the average value is transformed to typical value of $\{0, 1, 2, 3\}$ where 0 represents blank area, 1 represents an obstacle occupancy, 2 represents the robot's position, and 3 represents the appointed destination. A typical environment model is shown in Fig. 12, where all pixels in each grid are read and the average color component is obtained. The following grids are then read in the same manner. This goes on until the end of the 100th grid. Each grid is labeled by thresholding according to the ratio of the color component in them.

The actual bitmap image taken and the corresponding world map are shown in Fig. 4.13. Note that there is a virtual border of one grid unit on each side. The environment is an 8 by 8 grid units area. A one-grid length width is made '1' to be the border. Obstacles and walls are indicated by '1', start and goal positions are indicated by '2' and '3' respectively while empty area is indicated by '0'. The picture can be taken from any height provided that the camera's angle of focus is directly on the area concerned, perpendicular to the ground. No fixed distance between the camera and the world is therefore required. With the world grid map and position information from the encoders on its wheels, the robot computes its position in the world. If a new obstacle is found by its sensors, the value in the grid map will be updated for further planning.

4.4.2   GMRPL

As one of the reinforcement learning methods, the temporal difference (TD) learning [14] with replacing eligibility traces method has a significant improvement in learning performance and reduction of parameter sensitivity on specific application as compared to the conventional TD methods. The agent learns by comparing temporally successive predictions and it can learn before seeing the outcome. The agent maximizes the expected discounted return R_t over a certain number of time steps, called an episode, by summing all the rewards at each time step t,

$$R_t = r_{t+1} + \gamma r_{t+2} + \gamma^2 r_{t+3} + \ldots = \sum_{k=0}^{\infty} \gamma^k r_{t+k+1} \qquad (4.6)$$

where γ, $0 \leq \gamma \leq 1$, is the discount rate. The value function takes the following form,

$$V(s_t, a_t) \leftarrow V(s_t) + \alpha[r_{t+1} + \gamma V(s_{t+1}) - V(s_t)] \qquad (4.7)$$

where α is the time step and s_t denotes the state at time step t. This value function is extended to the Q-function where the expected return starting from that state, taking that action, and thereafter following the policy,

$$Q(s_t, a_t) \leftarrow Q(s_t, a_t) + \alpha[r_{t+1} + \gamma Q(s_{t+1}, a_{t+1}) - Q(s_t, a_t)]. \qquad (4.8)$$

If s_{t+1} is terminal, then $Q(s_{t+1}, a_{t+1}) = 0$.

This is the updating function of the Q table of the state-action pair. On each update, $Q(s_t, a_t)$ is an estimate from the previous $Q(s)$. In essence, the concept of temporal difference is the proportionality of error with the change over time of the prediction. It makes long-term predictions about dynamic systems. The SARSA(s, a, r, s', a') algorithm updates the estimation of the action-value functions $Q(s_t, a_t)$ at every time step [15].

The eligibility traces, $e_t(s)$ records temporarily the occurrence of an event, of either visiting of a state or taking of an action. Thus only eligible states or actions are assigned credit or blamed when a TD error occurs. In replacing traces, the frequently visited state trace is set to 1 while the others are decayed by a factor of $\gamma\lambda$

$$e_t(s) = \begin{cases} \gamma\lambda e_{t-1}(s) & \text{if} s \neq s_t \\ 1 & \text{if} s = s_t \end{cases}. \tag{4.9}$$

The SARSA(λ) algorithm is illustrated as follows:
Step 1. Initialize value function arbitrarily;
Step 2. Initialize the state s and action;
Step 3. Take action a, and observe r and s';
Step 4. Choose a' and s' following a certain policy;
Step 5. Update:

- $TDerror \leftarrow r + \gamma Q(s', a') - Q(s, a)$
- $e(s, a) \leftarrow 1$

where TDerror is the temporal difference error.

Step 6. Update for all s, a:

- $Q(s, a) \leftarrow Q(s, a) + \alpha\delta e(s, a)$
- $e(s, a) \leftarrow \gamma\lambda e(s, a)$
- $s \leftarrow s'; a \leftarrow a'$

where α is the step size and γ is the discount factor;

Step 7. Set $s = s'$;
Step 8. Repeat step 2 to 7 for a certain number of episodes.

The autonomous robot first gets the world model, learns its shortest path, and starts its motion with the first action at the first state. The action-state pair extraction cycle goes on until the goal position is reached. One recorded experiment trajectory of the Khepera using reinforcement learning is shown in Fig. 4.14. The experimental environment is constructed simulating an ordinary office room. The global camera obtains the space image, which can be transformed into a grid map. The state of the robot is determined by its coordinated value, and the optimal action at each state is selected by the reinforcement learning. As shown in Fig. 4.15, when there are new obstacles in the planned path, the robot makes adaptive modifications to the grid map and re-plan the optimal path during its navigation. In this figure, the circle point is the position at which the robot re-plans its path to the destined goal.

4.5   Real-Time Implementation

Several experiments have been conducted to examine the effectiveness of the autonomous robot controller in both virtual and real environment. The experimental configuration is shown in Fig. 4.16. The robot communicates with the host computer through a RS232 serial line. The controller program is written in C language and built in Visual C++ compilation environment. It can run in the host computer as well as to be downloaded onto the real robot. An example experiment environment is shown in Fig. 4.17. The experimental settings are described as follows:

- Target is defined as {light = 512; color = RGB(255, 0, 0); height = 58; resistivity = 255; size = 200}
- New objects similar to the target are set as, Object 1 = {light = 512; color = RGB(255, 0, 0); height = 28; resistivity = 255; size = 200};
  - Object 2 = {light = 100; color = RGB(255, 0, 0); height = 58; resistivity = 255; size = 200};
  - Object 3 = {light = 512; color = RGB(0, 0, 255); height = 58; resistivity = 255; size = 200};
  - Object 4 = {light = 512; color = RGB(0, 255, 0); height = 58; resistivity = 255; size = 200};
  - Object 5 = {light = 512; color = RGB(255, 0, 0); height = 58; resistivity = 255; size = 20};
- The experiment is conducted on an 80cm × 80cm black arena having a smooth and clear surface in order to prevent the robot from slipping or locking;
- Obstacles are distributed with standard distance among them so that the physical environment can be represented by a grid map. The obstacles' position may be changed during the execution of robot tasks;
- The destination for the robot is predefined.

In the experiment, the value obtained from the sensor is often not the exact value in the object's representation since the real world environment is often noisy and unstable. However, the recognition decision in our experiment is reliable in that the fuzzy logic is adopted to deal with the sensor data. Sensor fusion endows the robot the capability of recognizing target from other similar objects. When the environment is changed, the robot is capable of re-planning its path to destination with the reinforcement learning path planner. In our experiment, the intelligent strategies deployed in the projective agents proved to be effective in coping with uncertain environments. The controller model deployed for our robot has provided an easy-to-use framework in which intelligent sub-agents can be designed and integrated conveniently, which endows the robot the capability of increasing its intelligence with new agents added.

4.6   Summary

This chapter has reported our research on the development of fuzzy logic sensor data fusion and reinforcement learning for autonomous robot navigation. A feature-based multi-stage fuzzy logic sensor fusion system has been developed for target recognition, which is capable of mapping noisy sensor inputs into reliable decisions. The robot exploration and path planning are based on a grid map oriented reinforcement path learning system, which allows for long-term predictions and path adaptation via dynamic interactions with physical environments. These two intelligent modules enable an autonomous agent to accomplish the task of locating a targeted object among many similar objects and to move it to the destination. The real-world application using a Khepera robot shows the robustness and flexibility of the developed system in dealing with robotic behaviors such as target collecting in the ever-changing physical environment.

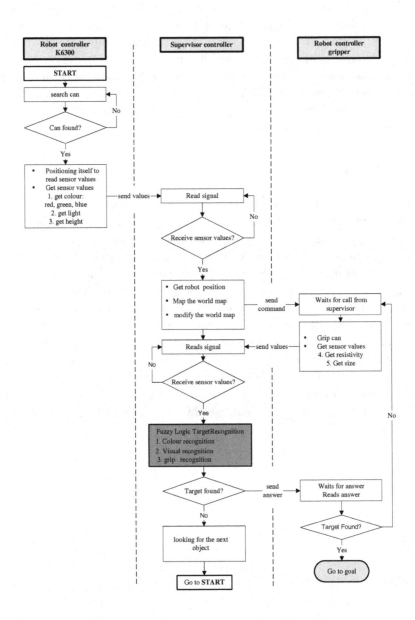

Fig. 4.7   Data flow diagram of the function modules.

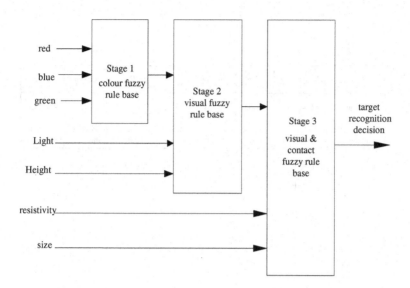

Fig. 4.8   Multi-stage fuzzy inference system.

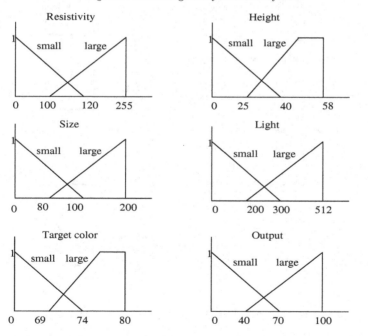

Fig. 4.9   The shape of the membership functions.

	Resistivity	Size	Height	Light	Red	Green	Blue
Range	$0-255$	$0-200$	$0-58$	$0-512$	0 - 255	$0-255$	0 - 255
Larger number means:	Smaller resistivity	Larger size	Larger height	Lower light	Higher intensity	Higher intensity	Higher intensity

Fig. 4.10   Range of the sensor inputs.

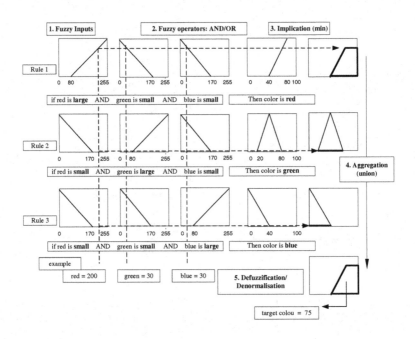

Fig. 4.11   The fuzzy logic inference system in the first stage.

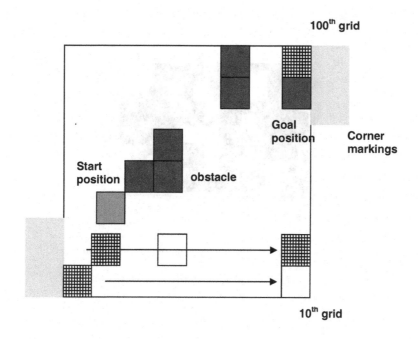

Fig. 4.12   Scanning the whole area and normalizing the region.

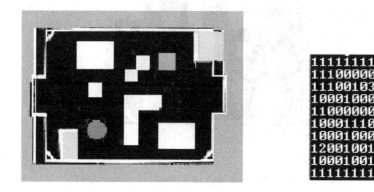

Fig. 4.13   Actual picture and its corresponding world map generated.

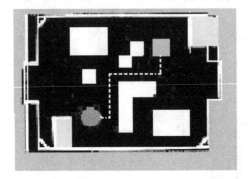

Fig. 4.14 Khepera path planning using reinforcement learning.

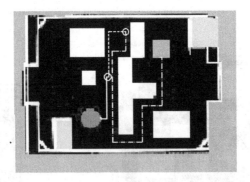

Fig. 4.15 Re-planning path in a changing environment.

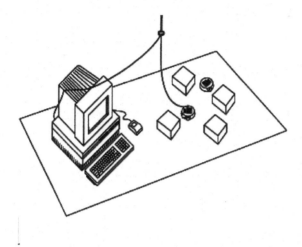

Fig. 4.16   The experimental configuration [16].

Fig. 4.17   Snapshot of the experimental environment.

Bibliography

[1] Luo, R. C., and Kay, M. G., Multisensor Integration and Fusion in Intelligent Systems, *IEEE Trans. on Systems, Man and Cybernetics*, vol. 19, no. 5, pp. 901–931, 1989.

[2] Delcroix, C. J. and Abidi, M. A., Fusion of range and intensity edge maps, in *Sensor Fusion: Spatial Reasoning and Scene Interpretation*, vol. 1003, pp 145–15, 1988.

[3] Mitiche, A. and Aggarwal, J. K., Image segmentation by conventional and information-integration techniques: a synopsis, *Image and Vision Computing*, vol. 3, no. 2, pp. 50–62, 1985.

[4] Russo, F., and Ramponi, G., Fuzzy Methods For Multisensor Data Fusion, *IEEE Trans. on Instrumentation and Measurement*, vol. 43, no. 1, pp. 288–294, 1994.

[5] Abdulghafour, M., Chandra, T., Abidi, M. A., Data fusion through fuzzy logic applied to feature extraction from multi-sensory images, *IEEE Trans. on Robotics and Automation*, vol. 2, pp. 359–366, 1993.

[6] Kam, M., Zhu, Xiaoxun, and Kalata, P., Sensor fusion for mobile robot navigation, *Proceedings of the IEEE*, vol. 85, Issue 1, pp. 108–119, 1997.

[7] Manyika, J., and Durrant-Whyte, H., *Data Fusion and Sensor Management: A Decentralized Information-theoretic Approach*, New York, 1994.

[8] Klir, G. J., and Yuan, B., *Fuzzy Sets and Fuzzy Logic: Theory and Application*, Prentice Hall, New Jersey, 1995.

[9] Cox, E., *The Fuzzy Systems Handbook: A Practitioner's Guide to Building, Using, and Maintaining Fuzzy Systems*, 2nd edition, AP Professional, Boston, 1998.

[10] Brooks, R. A., A robust layered control system for a mobile robot, *IEEE Journal of Robotics and Automation*, vol. 2, no. 1, pp. 14-23, 1986.

[11] Pfeifer, R., and Scheier, C., *Understanding Intelligence*, MIT Press, Cambridge, 1999.

[12] Sutton, R. S., and Barto, A. G., *Reinforcement Learning: An Introduction*, MIT Press, 1998.

[13] Yen, J., and Langari, R., *Fuzzy Logic: Intelligence, Control, and Information*, Prentice Hall, New Jersey, 1998.

[14] Sutton, R. S., Learning to predict by the methods of temporal difference, *Machine Learning*, vol. 3, pp. 9–44, 1988.

[15] Rummery, G., and Niranjan, M., *On-line Q-learning Connectionist Systems*, Tech. Rep. Technical Report CUED/F-INFENG/TR166, Cambridge, University engineering Department, 1994.

[16] K-Team Corp, *Khepera User Manual*, http://www.k-team.com.

[17] Cyberbotics Corp, *Webots User Guide*, Http://www.cyberbotics.com.

[18] Ayache, N., and Fangeras, O., Maintaining representations of the environment of a mobile robot, *IEEE Trans. Robotics and Automation*, vol.5, pp. 804–819, 1989.

[19] Mitchell, T. M., *Machine Learning*, The McGraw-Hill Companies, Inc., 1997.

[20] Conway, J. H., *Regular Algebra and Finite Machines*, Edschapman & Hall, 1971.

[21] Jennings, N. R., Sycara, K., and Wooldridge, M., A roadmap of agent research and development, *Autonomous Agents and Multi-Agent Systems*, vol. 1, pp. 7–38, Kluwer Academic, Boston, 1998.

[22] Wooldridge, M., and Jennings, N. R., Intelligent agents: theory and practices, *The Knowledge Review*, vol. 10, no. 2, pp. 26–37, 1997.

[23] Brooks, R. A., Intelligence without representation, *Artificial Intelligence*, vol. 47, no. 1-3, pp. 139–159, 1991.

[24] Varshney, P. K., Multisensor data fusion, *Electronics & Communication Engineering Journal*, pp. 245–253, 1997.

[25] Zadeh, L. A., Fuzzy sets, *Inform. Control*, vol. 8, pp. 338–353, 1965.

[26] Zadeh, L. A., Soft computing, fuzzy logic and recognition technology, in *Proc. FUZZ-IEEE, World Congress on Computational Intelligence*, Anchorage, AK, pp.1678–1679, 1998.

[27] Laird, J. E., Yager, E. S., Hucha, M., and Tuck, C. M., Robo-soar: An integration of external interaction, planning, and learning using Soar, *Robotics and Autonomous Systems*, vol. 8, no. 1–2, pp.113–129, 1991.

[28] Wooldridge, M., and Jennings, N. R., Intelligent agents: theory and practice, *The Knowledge Review*, vol. 10, no. 2, pp.115–152, 1995.

Chapter 5

Task-Oriented Developmental Learning for Humanoid Robots*

This chapter presents a new approach of task-oriented developmental learning for humanoid robotics. It is capable of setting up multiple tasks representation automatically using real-time experiences, which enables a robot to handle various tasks concurrently without the need of predefining the tasks. In this approach, an evolvable partitioned tree structure is used for the task representation knowledgebase. Since the knowledgebase is partitioned into different task domains, the search and updating of task knowledge only focus on a particular task branch, without considering the entire task knowledgebase which is often large and time-consuming in the process. A prototype of the proposed task-oriented developmental learning is designed and implemented using a Khepera robot. Experimental results show that the robot can redirect itself to new tasks through interactions with the environments, and a learned task can be easily updated to meet any varying specifications in the real world.

5.1 Introduction

One important characteristic of humanoid robots is the ability to learn varying tasks in changing human environments. Due to the varying nature of real world, it is impractical to predefine fixed task representation for a robot, which must be equipped with the capability of setting up task representation automatically through online experiences [1]-[2]. In addition, the robot should be able to handle multiple tasks concurrently with real-time responses. Many machine-learning algorithms have been proposed for ap-

---

*Portions reprinted, with permission, from (Tan, K. C., Chen, Y. J., Tan, K. K. and Lee, T. H., "Task-oriented developmental learning for humanoid robots", *IEEE Transactions on Industrial Electronics*, vol. 52, no. 3, pp. 906–914). ©2005 IEEE.

plications such as speech recognition [3],autonomous vehicle driving [4] and chess games playing [5]. These learning algorithms are often task-specific and working under the consideration of certain hypotheses [6]: Firstly, a set of pre-collected training samples should be available and it is crucial that the distribution of training samples should be identical to the distribution of test samples. Secondly, the learning task should be predefined in delicate representation. Finally, the world model should be given or partially given so that the robot can identify itself in the environment. However, these hypotheses cannot be assumed as a matter of course in typical human world, and sometimes they are even impossible to be realized. In fact, these hypotheses are the obstacles in developing practical robots working in human world that is often complex in nature.

These algorithms often need dedicated task representation, carefully collected training example and well-defined world model as the precondition. The perspective on these learning algorithms is that it involves searching a large space to determine the best behavior fitting the observed data and prior knowledge. When the task becomes more complicated or multiple, the search procedure is often incapable of satisfying the real-time responses. Although these algorithms are effective for specific tasks, they may not be able to deal with multiple tasks concurrently as required in humanoid robots. It is known that human brain can learn multiple tasks with online experiences simultaneously; the knowledge is evolved from interactions with the environment. Therefore the humanoid robot should be given the ability to organize its perception information into useful task knowledge when it is growing up, i.e., its task learning purpose is not predefined when it is born. Considering the gap between traditional learning algorithms and the humanoid robot's requirements, a new model called developmental humanoids has been proposed [7].

The approach of developmental humanoids [7] presents a new way to develop the robotics' task skills. The self-organizing schemes using clustering and classification tree [8]-[10] are adopted to guide the knowledgebase development that is able to deal with varying tasks depending on the inputs obtained from the robot's sensors. Comparing with traditional learning algorithms,developmental algorithm is able to learn varying tasks and no reprogramming is required for the developmental robot to adapt to new task learning. Although the developmental approach is capable of adapting to new tasks, its learning ability in specific task may be poorer than traditional learning algorithms [6]. Since the knowledge structure for this model is not task-oriented, the knowledge nodes for the same task are unrelated to each other in the knowledgebase. When the task becomes more complicated or the number of tasks is large, the required search time may be incapable of meeting the real-time requirements. Moreover, the new task in the developmental approach is often unrelated to previous learned task, even if the learned task is part of the new task. Considering this issue, an improved developmental learning system for humanoid robotics called task-oriented developmental learning (TODL) is proposed in this chapter.

In our approach, the knowledgebase is organized differently from the non-task specific developmental learning algorithm HDR [7], i.e., the knowledge node for the same task purpose is interrelated to each other and is integrated into one task module in the knowledgebase. The task module is determined by general task information, such as task purpose, task status, which is keyed in by the human trainer via a human-machine GUI. Inside each task module, the knowledge nodes of sensors' information and actuators' status are organized into a self-evolving cluster and classification tree. A friendly human-machine GUI is developed to facilitate interactions between the robot and human trainer instead of any specific sensors. The proposed learning system offers a number of advantages: The search time for a piece of knowledge on certain task purpose is much shorter than the non-task specific developmental model. The advantage becomes more obvious for complex tasks or when the number of tasks is large. With the proposed task-oriented developmental structure, a robot can handle multiple tasks concurrently without the need of predefining any tasks. The task representation is defined by state space which is the key point where the robot takes certain actions. It is represented by sensor data obtained from external sensor mounted on the robot like camera and infrared sensor, and the action is the command given by the human trainer such as motor speed and move direction. All these information is obtained from real-time experiences and is organized by the learning algorithm into an efficient data structure.

The following sections are organized as follows: Section 5.2 presents the general structure of the proposed task-oriented developmental learning (TODL) system. Section 5.3 describes the learning algorithm for self-organized knowledgebase (SOKB) and Section 5.4 presents a prototype implementation of the TODL system. A task-learning experiment for the proposed TODL system is shown in Section 5.5. Discussions for the results are given in Section 5.6 and conclusions are drawn in Section 5.7.

5.2   Task-Oriented Developmental Learning System

5.2.1   *Task Representation*

In task-oriented developmental learning system, a robot learns and builds up its knowledgebase while it is performing the tasks. Unlike traditional approaches, the knowledge in TODL is stored and retrieved by the robot concurrently. The task representation is constructed as a mapping between the robot's situation space and the suitable action space. The situation space composes of different states requiring an action to be made by the robot. The state and action pair is integrated into a record, which forms a knowledge node on the knowledge tree. The knowledge retrieval process is to map the current state to the situation that the robot has remembered in its knowledgebase as well as to retrieve the corresponding action it takes under such a situation. A state is determined by three kinds of information in the task-oriented approach, i.e., sensor perception, actuator status and task description. The sensor perception consists of both stimuli from external environment, e.g., visual and proximity, and relative position of internal control, e.g., gripper position. The actuator status refers to the last performed action, e.g., the hand is empty or is carrying objects, and the task description labels the robot's task purpose under the current situation, e.g., task label and task step. The task-oriented developmental learning can be termed TOAA-learning (task-oriented automated animal-like learning), which is based on the AA-learning architecture [7] but includes the task information in the learning operation.

5.2.2   *The AA-Learning*

The robot agent M conducts AA-learning [7] at discrete time instances, $t = 0, 1, 2, \ldots$, if the following conditions are met: (a) M has a number of sensors including the external and internal sensors whose signal at time t is collectively denoted by $x(t)$; (b) M has a number of effectors whose control command at time t is collectively denoted by $a(t)$; (c) The task purpose at time t is denoted by $T(t)$; (d) M has a "brain" denoted by $b(t)$ at time t; (e) At each time t, the time-varying state-update function f_t updates the "brain" based on sensory input $x(t)$, task purpose $T(t)$ and the current "brain" $b(t)$

$$b(t + 1) = f_t(x(t), T(t), b(t)).\qquad(5.1)$$

The action-generation function g_t generates the effector's control command based on the updated "brain" $b(t+1)$

$$a(t+1) = g_t(b(t+1)) \qquad (5.2)$$

where $a(t+1)$ can be part of the next sensory input $x(t+1)$; (f) the "brain" of M is closed in that after the birth, and $b(t)$ cannot be altered directly by human trainer for teaching purpose. It can only be updated according to equation (5.1).

5.2.3   Task Partition

In the proposed TODL system, the "brain" $b(t)$ is partitioned into different task modules based on the task memory $TM(t) = (T(t-1), T(t-2), \ldots, T(1))$ using classification tree, and each task module is a self-organized hierarchical discriminant tree based on previous sensory input vectors $XM(t) = (x(t-1), x(t-2), \ldots, x(t-k))$, where k is the temporal extent of sensory inputs. These input vectors are organized into different clusters based on their relativity determined by the distances among each others. The "brain" is updated by the function f_t. The generation of action $a(t+1)$ with g_t is realized through a mapping engine that accepts current $(x(t), T(t))$ and previous memory of $(TM(t), XM(t))$ as the input and generates $a(t+1)$ as the output. The $x(t)$ and $T(t)$ also updates the long term memory of $b(t+1)$ for each time instance of t.

The action $a(t)$ is a member of the predefined action set $A = \{A_1, A_2, \ldots, A_n\}$. The actions in the action set can be divided into two groups: One group is called positive action, which is carried out by a robot when no trainer is available, and the robot will improve its experience through a trial-and-error process. The other group is named passive action, which is received from the trainer's teaching. The trainer can also provide performance assessment to positive behaviors evolved from the robot, if necessary. The action set will remain unchanged after the robot is 'born' and the set is programmed as an interface between the front-end robot and the back-end knowledgebase. When different robots are employed in the TODL system, only the interface is required to be rewritten without affecting the learning algorithm and the knowledgebase. Although it is possible for a robot to learn new actions through life-span learning, such learning is often too time-consuming and impractical. Intuitively, tasks can be regarded as complicated actions, and by giving the robot a predefined action set, it will help to facilitate the learning process. In the proposed TODL system, passive action has a higher priority than positive action during the learning process. When no action command is received from the human trainer, the robot will explore the surrounding world with its positive actions.

5.3   Self-Organized Knowledgebase

5.3.1   *PHDR Algorithm*

A learning system involves the search of possible hypotheses in a large space or knowledgebase in order to find one that best fits the observed data [6]. In traditional knowledgebase development [11], the search space is predefined for the robot considering the specific task purpose and environment, and the learning process is to set up rules of choosing the best action from the legal action sets under a defined situation. However, this approach is not suitable for intelligent robot working in humanoid world, for which the robot has no specific task purpose or idea about the working environment when it is 'born'. Therefore the robot should be capable of defining the search space by itself through past experiences and choosing the optimal action under the learned situation. In designing a knowledgebase for humanoid robot, we adopt the approach of self-organized knowledgebase (SOKB) that allows a robot to construct and update the search space through on-line experiences. The learning method for SOKB is called partitioned hierarchical discriminant tree (PHDR), which combines the traditional classification tree method CART [8] with the new hierarchical discriminant regression (HDR) algorithm [9].

5.3.2   *Classification Tree and HDR*

Classification tree analysis [12] is one of the main techniques used in predicting membership of cases or objects in the classes of a categorical variable through measuring predictor variables. The process of organizing a traditional classification tree involves four basic steps: (1) Specifying the criteria for predictive accuracy; (2) Selecting splitting rules; (3) Determining the stop of splitting; and (4) Pruning the tree to the optimal size. The methodology for each step has been widely discussed in [8], [13]-[14]. In our SOKB system, camera image is used as the main source of sensory input since image can afford a large amount of information for a robot to determine the situations. However, image recognition is a difficult task particularly in real environments. The traditional content-based image retrieval [15]-[16] used manually predefined features as the image classifiers, which is difficult to be generalized for our developmental model with a large and changing task domain. Therefore the appearance-based image retrieval [17]-[19] is employed in the SOKB to enable a robot to automatically derive features from image samples. In this approach, the two-dimensional image is represented as a long vector, and statistical classification tools are applied to the sample vectors directly. However, traditional decision trees for low-dimensional feature space [8], [20]-[21] are not suitable for input dimensionality of a few thousands or larger, as encountered in our developmental approach.

Recently, a new method called hierarchical discriminant regression (HDR) [9] is proposed to classify high-dimensional sensory input data. In HDR, the tree grows from the root node where the input samples are assigned into q clusters based on the nearest Euclidean distance. Each cluster is represented by its mean of samples belonging to it as its center and the covariance matrix as its size. If the size of a cluster is larger than the defined threshold, then this cluster will be divided into another q clusters as the root node until a cluster with defined size is obtained, and this cluster will be taken as a leaf node. In HDR, the deeper a node is in the tree, the smaller the variance of its cluster is. The center of each cluster provides essential information for discriminating subspaces. A total of q centers of q clusters at each node give $(q-1)$ discriminating features, which span $(q-1)$ dimensional discriminating space. With HDR, the discriminating features need not be predefined but can be derived from the input samples. Therefore the high-dimensionality classification task is translated into $(q-1)$ dimensionality classification problem. As discussed in Section 5.2, however, HDR can grow very large when the number of learned tasks becomes large, which could be very time-consuming in meeting the real-time requirements since the search for a certain task needs to go through the entire tree. To address the issue, a new classification tree in the knowledgebase self-organization called partitioned hierarchical discriminant regression tree (PHDR) based on the HDR [9] is proposed. In this approach, the input samples are first classified using classification tree with task description as the discriminating feature, e.g., the input space will be partitioned into different task domain. Inside each domain, the HDR is then used to classify the high-dimension sensory input.

5.3.3 *PHDR*

In SOKB, the knowledgebase accepts a set of on-line training samples (x_i, y_i) as input, while $x_i \in X$ and $y_i \in Y$, and generates an approximate mapping $X \rightarrow Y$ as output. X is the input space constructed by sensory data collections (e.g., image, infrared) and task descriptions, while Y is the output space of the legal action set for a robot. The mapping is updated for every new sample (x_i, y_i). With PHDR, the large X space is first partitioned into smaller X' space labeled by the task descriptions, and each X' space is then applied LDA to self-organize a task-specific classification tree. Since the robot holds only one task purpose at one time, the search should be performed in the task-specific partitioned space in order to lessen the searching time. Another advantage of the partitioned approach is that only one partitioned classification tree is updated when a new sample (x_i, y_i) arrives, which relieves the updating work. In our approach, searching and updating is running in parallel. The front-end robot agent only awaits a search process to end and the updating work is left to the back-end database to finish before the next search command comes in. The partitioned procedure is described as follows,

Given training sample set $L = \{(x_i, y_i) | i = 1, 2, \ldots, n, x_i \in X, y_i \in Y\}$, x_i is a long vector combined with two vectors S_i and T_i, while S_i is the sensory input (e.g., camera, infrared), and T_i is the task description corresponding to S_i. Then L is divided into smaller set $L'(T_p) = \{(s_k, y_k) |$ while $T_k = T_p; T_p \in T; T$ is the collection of learned task domains$\}$.

In a partitioned space X', the knowledge is task-specific and HDR is employed to incrementally build a classification tree from a sequence of training samples. A tree has the characteristics of automatic features deriving and fast in searching time. Given training sample set $X' = \{(s_i, y_i) | i = 1, 2, \ldots, m, s_i \in S, y_i \in Y\}$, the task is to determine the class label of any unknown input $s \in S$. The self-evolving procedure of HDR [9] is described as follows,

Build Tree: Given a node N and a subset X' of the training samples that belong to N, $X' = \{(s_i, a_i) | s_i \in S, a_i \in A, i = 1, 2, \ldots, m\}$, build the tree which roots from the node N using X' recursively. At most q clusters are allowed in one node:

- Let p be the number of the s clusters in node N. Let the mean C_1 of s-cluster 1 be s_1, Set $p = 1$ and $i = 2$.
- For $i = 2$ to n do For $j = 1$ to p do

- Compute the Euclidean distance $d = dist(s_i, C_j)$, where C_j is the jth s-cluster mean
- Find the nearest s-cluster j for s_i with the smallest d.
- If $d >= \delta$ and $p < q$, let the mean $C_p + 1$ of s-cluster $p + 1$ be s_i. Set $p = p + 1$.
- Otherwise, assign the sample (s_i, y_i) to cluster j. Compute the new mean C_j of jth s-cluster.

• For each s-cluster in node N, we now have a portion of samples S_j assigned to it and the mean of these samples S_j is C_j. Let k_j be the samples number inside jth cluster.
For $j = 1$ to q do

- For $l = 1$ to k_j do
  Compute the Euclidean distance $d = dist(s_l, C_j)$, where C_j is the jth s-cluster mean
- Find the largest d, if this largest Euclidean distance is larger than δ, a child node N_j of N is created from the s-cluster and this build-tree procedure is called recursively with input amples S_j and node N_j. δ is a valve value number to determine whether two s sample is approximately the same. This number represents the sensitivity of the classification tree.

This procedure creates a classification tree with root R and all the training samples. The procedure to query the classification tree for an unknown sample s is described as follows:

Retrieval Tree: Given a classification tree T and a sample s, find the corresponding action a. The parameter k specifies the upper bound in the width of parallel tree search.

(1) From the root of the tree, compute the probability-based distance to every cluster in the node. Select at most top k s-clusters, which have the smallest probability-based distances to s. These s-clusters are called active s-clusters.
(2) For every active cluster received, check if it points to a child node. If it does, mark it inactive and explore its child node as in Step 1. At most $k * k$ active s-clusters can be returned.
(3) Mark at most k active s-cluster according to the smallest probability-based distances.
(4) Do the above steps 2 through 3 recursively until all the resulting active s-clusters are all leaf node of the tree.

(5) Let the cluster c has the shortest distance among all active leaf nodes and the cluster mean is C. If $d = \text{distance}(s, C) < \delta$, output the corresponding a. Otherwise, the trainer will be asked to afford a new training sample (s, a) and this sample will be used to update the tree using the build tree procedure on the leaf node c.

(6) If this a is not suitable to be assessed by the human trainer, then a new a received from the trainer will be used to update the label of the selected leaf node.

5.3.4 *Amnesic Average*

The developmental learning is a life-span learning process where new input samples arrive continuously as the robot is performing. The initial center of each state cluster is largely determined by the early input data. When more data are available, these centers move to more appropriate locations. If these new locations of the cluster centers are used to judge the boundary of each cluster, the initial input data were typically classified incorrectly. In other words, the center of each cluster contains some earlier data that do not belong to this cluster. To reduce the effect of these earlier data, the amnesic average can be used to compute the center of each cluster. The amnesic average [7] can also track dynamic changer of the input environment better than a conventional average. The average of n input data $x_1, x_2, \ldots, x_n$ can be recursively computed from the current input data x_n and the previous average X_{n-1} by equation (5.3),

$$X_n = \frac{(n-1)X_{n-1} + x_n}{n} \tag{5.3}$$

i.e., the previous average X_{n-1} gets a weight $\frac{(n-1)}{n}$ and the new input $X(n)$ gets a weight $\frac{1}{n}$. These two weights sum to the value of 1. The recursive equation (5.3) gives an equally weighted average. In amnesic average, the new input gets more weight (determined by the factor l)than old inputs as given by,

$$X_n = \frac{(n-l)X_{n-1} + l * x_n}{n}. \tag{5.4}$$

5.4   A Prototype TODL System

As shown in Fig. 5.1, a prototype has been built to demonstrate the reliability and flexibility of the proposed TODL system. The structure of the system is shown in Fig. 5.2, which is divided into three modules: working robot agent, TODL mapping engine and knowledge database. These three modules are integrated with two interfaces, e.g., interface 1 connects the robot with mapping engine, and interface 2 connects the mapping engine with knowledge database. The two interfaces act as the middle-tier among these three modules such that any changes inside one module will not require the change of other modules. For example, when a robot from different vendor is employed, the TODL mapping engine and knowledge database need not be changed.

Fig. 5.1   A prototype TODL system.

5.4.1   *Robot Agent*

5.4.1.1   *Khepera robot*

The Khepera robot [22] is employed as the base platform in our implementation, where additional function modules such as gripper can be plugged on top of the base robot. The Khepera is equipped with eight IR sensors, six in the front and two in the back, which can return the proximity between surrounding and the robot. It has two wheels which are individually

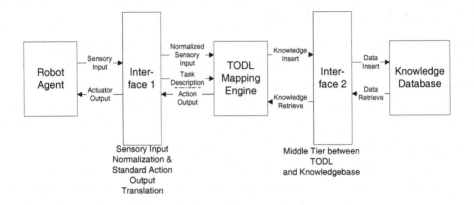

Fig. 5.2   The structure of the TODL system.

driven by DC motors. The speed of each wheel can be controlled through a PID controller, and on each wheel an incremental encoder is placed to calculate the wheels' position. A digital video camera is amounted on top of the robot to capture environment's images and a gripper is used with the robot to manipulate objects. The Khepera robot communicates with the host computer through serial port, while the video camera communicates with the host computer through USB port. The host computer is a PIII 933MHZ PC with 512MB RAM for real-time sensory information processing, real-time knowledge retrieval and update, as well as real-time actuator's controls in the implementation.

5.4.1.2   *Interface*

The interface 1 is designed for two purposes, i.e., for communication between the robot agent and TODL mapping engine, as well as for communication between the human trainer and TODL mapping engine. The GUI of interface 1 is shown in Fig. 5.3. The TODL system first receives the task domain and task step information from human trainer to locate the knowledge subspace, and the sensory data from the robot is then normalized and used as input to the TODL mapping engine to determine its situation. The sensory inputs are represented as a long vector $s =$ (Image, IR0~7, Rightmotorspeed, Leftmotorspeed, gripperarmposition, gripperhandStatus). If the mapping engine finds an approximate experience of the current situation, the corresponding action will be retrieved. Then the action command, which is a member of the standard legal action set, is translated by interface 1 to the actuator's controls for a particular hardware. The legal action set used in our experiment is $A = \{$turnleft, turnright, turnback, search, catch, sendtogoal$\}$. An innate action set is designed for the robot to learn by itself through trial-and-error, $A' = \{$avoid, follow, push$\}$. For different robots, the implementation of an action could be different due to different hardware particulars. However, through interface 1, the hardware particulars of the robot agent is not a concern for the TODL mapping engine in our implementation. The ODBC middleware is used for interface 2, which connects the mapping engine with knowledge database. With this middleware, the mapping engine can work independently of the database in use, as desired.

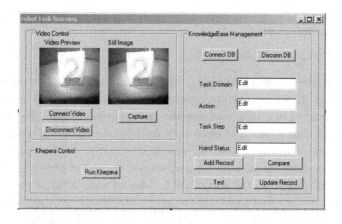

Fig. 5.3   Interface 1 between robot agent and TODL mapping engine.

5.4.2   *TODL Mapping Engine*

The value of q is set to 2 in the PHDR algorithm and the knowledge is represented in two tables, e.g., one for tree indexing and the other for leaf node. The tree-indexing table is organized as (child1id, child2id, nodecenter, nodesnumber). While child1id and child2id point to the child node of the current node, the id is assigned by the database automatically as an 18-byte number. Nodecenter is the mean of all the sensory input samples and nodesnumber is the number of samples inside this node. The leaf node of the knowledge tree is represented as (image, taskdomain, taskstep, action, handstatus, IR data, armposition, leftspeed, rightspeed). The leaf node is located through the PHDR algorithm and the corresponding action will be retrieved for the control of the robot.

5.4.3 *Knowledge Database*

The commercial database of Oracle8i [23] is adopted to manage the knowledgebase in our approach since the Oracle8i software has afforded much functionality to manage the database, especially the 8i edition has a data type called BLOB which can be used for image data. Furthermore, this software supports the necessary functionalities in developing a distributed learning system for our future work.

5.4.4 *Sample Task*

5.4.4.1 *Goods distribution problem*

Goods distribution is a promising task for humanoid robot in commercial applications. However, current robotic application in this area is often limited to fixed environment and specific task representation. To generalize the robot to learn multiple tasks in a flexible and changing environment, the following challenges need to be solved: Firstly, goods representation should be learned through on-line experience instead of pre-definition. Before the learning task, the robot has no idea of what kind of goods it will manipulate. The robot must able to set up the goods representation automatically through online sensory information. Secondly, the task sequence should be learned through interactions with environment instead of predefined representation. When the goods distribution sequence is changed, the robot should be able to adapt to new task without reprogramming. In this implementation, the robot is to learn a sample goods distribution task as shown in Fig. 5.4 using the proposed TODL approach.

5.4.4.2 *Objects identification*

In the task shown in Fig. 5.4, the robot needs to manipulate 5 kinds of goods labeled with digit number of 1, 2, 3, 4, and 5. Each kind of goods may have multiple instances which are scattered around an experimental environment. The robot needs to collect these goods in a specified sequence from 1 to 5 to a target place identified by digit number 6. A problem in implementing this task is to identify the objects without being given predefined representation. In our experiment, camera image is used as the main source of information in order to determine the situations. Besides, infrared sensor is used to determine the distance between the robots to the goods to ensure that the image taken is in similar scale at different time. Since the objects to be manipulated by the robot are neither predefined nor fixed, the appearance-based image retrieval approach [17]-[19] is employed to enable the robot to automatically derive features from the on-line image samples as described in Section 5.3.2.

5.4.4.3 *PHDR representation of the sample task*

The knowledgebase accepts a set of on-line training samples, the input X space constructed by the sensory data collections (image, infrared) and task descriptions from human interface, while the output Y space composes of the legal action set for the robot. With PHDR, the large X space is first partitioned into smaller X' space labeled by the task descriptions, and each X' space is then applied LDA to self-organize a task-specific classification tree as described in Section 5.3.3. The sample images of the 5 objects and the goal place are shown in Fig. 5.5. The first-step in the whole task of collecting goods 1 can be represented in the robot's knowledgebase as shown in Table 5.1.

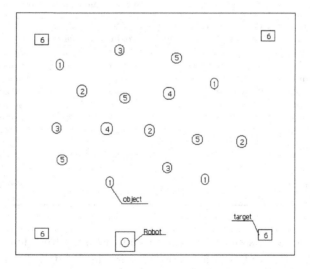

Fig. 5.4   Sample task: Collect scattered objects into target place at specified sequence.

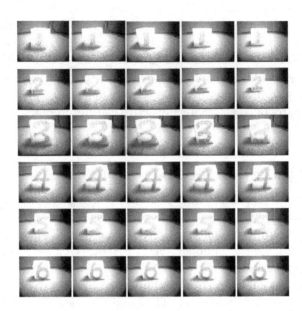

Fig. 5.5   Online object images captured by the camera on the robot.

Table 5.1   First-step representation in the sample task.

Task domain	Task step	Hand status	Object image	IR proximity	Other sensors	Action
Sample	0	0	1	near	. . .	**catch**
Sample	0	0	2	near	. . .	avoid
Sample	0	0	3	near	. . .	avoid
Sample	0	0	4	near	. . .	avoid
Sample	0	0	5	near	. . .	avoid
Sample	0	0	6	near	. . .	avoid
Sample	0	1	1	near	. . .	avoid
Sample	0	1	2	near	. . .	avoid
Sample	0	1	3	near	. . .	avoid
Sample	0	1	4	near	. . .	avoid
Sample	0	1	5	near	. . .	avoid
Sample	0	1	6	near	. . .	**release**

The robot learns the task domain and task step from human trainer through the designed GUI and obtains other information from its sensor and actuator status. The steps of collecting goods from 2 to 5 are similar to the first-step. If a new task sequence is required, the robot can easily adapt to the new requirement through online experiences. It can modify the old task domain or construct a new task domain to suit the new requirements. For example, Table 5.2 shows a modified first-step representation for the new requirement of collecting goods from the sequence of 5 to 1.

Table 5.2   Modified first-step representation in the sample task.

Task domain	Task step	Hand status	Object image	IR proximity	Other sensors	Action
Sample	0	0	1	near	...	avoid
Sample	0	0	2	near	...	avoid
Sample	0	0	3	near	...	avoid
Sample	0	0	4	near	...	avoid
Sample	0	0	5	near	...	**catch**
Sample	0	0	6	near	...	avoid
Sample	0	1	1	near	...	avoid
Sample	0	1	2	near	...	avoid
Sample	0	1	3	near	...	avoid
Sample	0	1	4	near	...	avoid
Sample	0	1	5	near	...	avoid
Sample	0	1	6	near	...	**release**

The challenge of approximating the image and other sensor's high dimensional input data in real-time has been discussed in previous sections. The implementation results show that the robot is capable of learning the task successfully. When the robot needs to manipulate other objects not labeled with digit number, such as apple or orange, a new task domain can be easily opened for the new requirements.

5.5   Discussions

With the proposed task-oriented developmental learning (TODL) system, a robot can develop its mental skills for multiple tasks automatically through real-time interactions with the environment. The disorderly and unsystematic sensory information is organized into structured knowledge about certain task purpose through the TODL system under real-time requirements. A technical challenge of task learning methodologies for intelligent robotics is that the learning methods must be scalable as well as efficient. The tradeoff between the defined representation and the flexibility is very boring. The more knowledge about the task and working environment is given to the robot, the less flexibility the robot will obtain to adapt to a new environment and new task. In contrast, the more flexibility the robot is endowed with, the less efficient the robot is in learning a specific task.

The proposed TODL system marks an important technical advancement in the task learning of a robot, which has the efficiency in specific task learning as well as the scalability to general task purpose. In our TODL system, no predefined representation of the environment or task purpose is required for the learning of a robot. The TODL extracts useful information about certain task from its experience and generates the knowledge search space automatically. Therefore the boring problem of defining an arbitrary object from arbitrary background for arbitrary task is solved. The knowledgebase is organized as task-oriented trees in our TODL system. This approach allows effective and specific task learning algorithms, such as speech recognition or task planning, to be employed in the TODL system for specific purpose. Furthermore, the knowledge space is partition into specific task field in the TODL system, which allows the search time to match the real-time requirements in a real world environment.

5.6   Summary

A new task oriented developmental learning approach for humanoid robotics has been presented in this chapter, which allows the robot to learn multiple tasks in unknown environment without predefined task representation. In this approach, the task representation can be set up automatically through on-line and real-time interactions with the environment. The knowledgebase has a self-organized partitioned tree structure, and the partition is based on general task description where each particular task domain is a self-evolving classification tree. Since the knowledgebase is task oriented, the search and update of task knowledge only affect a particular task domain, which reduces the computation time needed in the learning. A prototype of the proposed task-oriented developmental learning system has been implemented. The experimental results show that the robot can redirect itself to new tasks through interactions with the environment, and a learned task can be easily updated to meet the changing requirements.

The efficiency of the system depends very much on the strength of basic function modules, such as the preprocessor of visual information and the innate behavior system. Currently only the appearance-based image processing for vision analysis has been deployed, and the navigation system has no high-level path planning function. However, these basic function modules are the basis of high-level task learning. A complicated task not only depends on effective knowledgebase architecture, but also the efficiency of sensors and actuators. Future work will be focused on performance improvement through better sensor module and innate behaviors.

Bibliography

[1] J. Weng, J. McClelland, A. Pentland, O. Sporns, I. Stockman, M. Sur and E. Thelen, Autonomous mental development by robots and animals, *Science*, vol. 291, no. 5504, pp. 599 – 600, 2000.

[2] S. Thrun, J. O'Sullivan. Clustering learning tasks and the selective cross-transfer of knowledge, *Learning To Learn*, S. Thrun and L. Pratt, ed., Kluwer Academic Publishers, 1998.

[3] K. Lee, *Automatic Speech Recognition: The Development of the Sphinx System.* Boston: Kluwer Academic Publishers. 1989.

[4] D. A. Pomerleau, *ALVINN: An autonomous land vehicle in a neural network.* (Technical Report CMU-CS-89-107), Pittsburgh, PA: Carnegie Mellon University.

[5] G. Tesauro, Practical issues in temporal difference learning, *Machine Learning*, vol. 8, no. 257, 1992.

[6] T. M. Mitchell. *Machine Learning.* MIT Press and the McGraw-Hill Companies, Inc. 1997.

[7] J. Weng, W. S. Hwang, Y. Zhang, C. Yang and R. Smith, Developmental humanoids: Humanoids that develop skills automatically, *Proc. the first IEEE-RAS International Conference on Humanoid Robots*, Cambridge, MIT, Sep. 7–8, 2000.

[8] L. Breiman, *Classification and regression trees*, Belmont, Calif. Wadsworth International Group, c1984.

[9] W. Hwang and J. Weng, Hierarchical discriminant regression, *IEEE Trans. Pattern Analysis and Machine Intelligence*, vol. 22, no. 11, pp. 1277–1293, November 2000.

[10] B. S. Everitt, S. Landay and M. Leese, *Cluster Analysis*, New York, Oxford University Press, 2001. 4th ed.

[11] K. B. Yoon, *A Constraint Model of Space Planning.* Boston, Computational Mechanics Publications, c1992.

[12] A. Webb, *Statistical Pattern Recognition*, London: Arnold, 1999.

[13] B. D. Ripley, *Pattern Recognition and Neural Networks.* Cambridge University Press, January 1996.

[14] W. Y. Loh and Y. S. Shih, Split selection methods for classification trees,

Statistica Sinica, vol. 7, no. 4, October 1997.

[15] L. G. Shapiro and G. C. Stockman, *Computer Vision*, Upper Saddle River, NJ: Prentice Hall, 2001.

[16] B. Furht, *Handbook of Multimedia Computing*, Boca Raton, FL: CRC Press LLC, 1999.

[17] I. Ulrich and I. Nourbakhsh, Appearance-based place recognition for topological localization, *Proceedings of ICRA 2000*, vol. 2, pp. 1023–1029, April 2000.

[18] H. Murase and S. K. Nayar, Visual learning and recognition of 3D objects from appearance, *International Journal of Computer Vision*, vol. 14, no. 1, pp. 5–24, January 1995.

[19] S. K. Nayar, S. Nene and H. Murase, Real-time 100 object recognition systems, *Proc. of the IEEE International Conference on Robotics and Automation*, Minneapolis, vol. 3, pp. 2321–2325, April 1996.

[20] S. K. Murthy, Automatic construction of decision trees from data: A multidisciplinary survey, *Data Mining and Knowledge Discovery*, vol. 2, no. 4, pp. 345–389, 1998.

[21] S. R. Safavin and D. Landgrebe, A survey of decision tree classifier methodology, *IEEE Trans. Systems, Man and Cybernetics*, vol. 21, no. 3 pp. 660–674, May/June 1991.

[22] K-Team. Corp., *Khepera User Manual*, (http://www.k-team.com).

[23] Oracle Corp., (http://www.oracle.com)

Chapter 6

Bipedal Walking Through Reinforcement Learning*

This chapter presents a general control architecture for bipedal walking which is based on a divide-and-conquer approach. Based on the architecture, the sagittal-plane motion-control algorithm is formulated using a control approach known as Virtual Model Control. A reinforcement learning algorithm is used to learn the key parameter of the swing leg control task so that stable walking can be achieved. The control algorithm is applied to two simulated bipedal robots. The simulation studies demonstrate that the local speed control mechanism based on the stance ankle is effective in reducing the learning time. The algorithm is also demonstrated to be general in that it is applicable across bipedal robots that have different length and mass parameters.

6.1 Introduction

It is a great challenge for scientists and engineers to build a bipedal robot that can have the similar agility or mobility of a human being. The complexity of bipedal robot control is mainly due to the nonlinear dynamics, unknown environment interaction and limited torque at the stance ankle.

Many algorithms have been proposed for the bipedal walking task [1]–[9]. To reduce the complexity of the bipedal walking analysis, some of the researchers restricted themselves to planar motion studies, some adopted simple models or linearization approaches, etc.

Honda humanoid robots (P2 and P3) [6] are the state-of-the-art 3D bipedal walking systems. The control method is based on playing back

*The original publication (Chew, C.-M. and Pratt, G. A., "Dynamic bipedal walking assisted by learning", *Robotica*, vol. 20, pp. 477–491.) is acknowledged. Copyright (2002) Cambridge University Press. Reprinted with permission.

(with modulation) trajectory recordings of human walking on different terrains [10]. Though the resulting walking execution is very impressive, such a reverse engineering approach requires iterative tunings and tedious data adaptation. These are due to the fundamental differences between the robots and their human counterparts, for example, the actuator behaviors, inertia, dimensions, etc.

MIT Leg Laboratory has designed and built a 3D bipedal robot called M2. This study investigates a control approach for M2 to achieve dynamic walking behavior. The control architecture is based on a divide-and-conquer framework. Through the framework, the walking task is decomposed into smaller subtasks. A control algorithm which is composed of a number of sub-algorithms can then be formulated. Reinforcement learning methods are applied to those subtasks that do not have simple solutions.

Several researchers [11]–[14] have proposed various learning-based algorithms for bipedal locomotion. Benbrahim and Franklin [15] have applied reinforcement learning for a planar biped to achieve dynamic walking. They adopted a "melting pot" and modular approach in which a central controller used the experience of other peripheral controllers to learn an average control policy. The central controller and some of the peripheral controllers used simple neural networks for information storage. One disadvantage of the approach was that nominal joint trajectories were required to train the central controller before any learning began. This information is usually not easily obtainable.

In our approach, no nominal trajectories of the joints are required before the learning takes place. Since we apply learning only to smaller subtasks, the learning processes are less complex and more predictable. The learning part plays a complementary role in the overall control algorithm.

The scope of this chapter covers the sagittal-plane motion-control algorithm for the biped to walk on a level ground and along a linear path. The analysis will be carried out in a dynamic simulation environment. Section 6.2 first describes two bipedal robots whose simulation models are used to validate the control algorithms developed in this research. Section 6.3 describes the proposed control architecture based on which the control algorithms for the bipedal walking task are formulated. Section 6.4 introduces two implementation tools that are adopted for the control algorithm. Section 6.5 describes the control algorithm for the sagittal-plane motion. Section 6.6 describes two simulation studies which are used to illustrate different characteristics of the control algorithm.

6.2   The Bipedal Systems

Two bipedal robots are considered in this chapter. One of them is a 7-link planar bipedal robot called Spring Flamingo (see Figure 6.1). It is constrained to move in the sagittal plane. The total weight is about 12 kg. The legs are much lighter than the body. Each leg has three actuated rotary joints. The joint axes are perpendicular to the sagittal plane. The biped has a rotary angular position sensor at each DOF. It is used to measure the relative angular position between consecutive links. There is also a sensor to detect the pitch angle of the body. Each foot has two contact sensors at the bottom, one at the heel and one at the toe, to determine ground contact events.

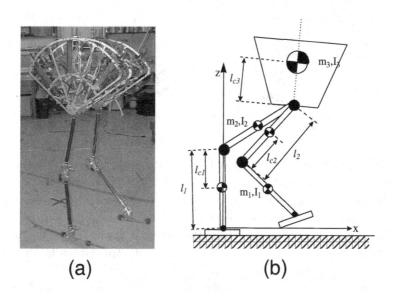

(a)   **(b)**

Fig. 6.1   7-Link 6 DOF planar biped - Spring Flamingo.

The other robot is a 3D biped called M2 (Figure 6.2). The total weight is about $25kg$. It has more massive legs compared with Spring Flamingo. Each leg has six active DOF of which three DOF is available at the hip (yaw, roll, pitch), one at the knee (pitch) and two at the ankle joint (pitch, roll). Each DOF has an angular position sensor to measure the relative angle between two adjacent links. Each of the feet has four force sensors (two at the toe and two at the heel) to provide the contact forces between the feet and the ground. Gyroscopic attitude sensor is mounted in the body to detect its roll, pitch and yaw angles.

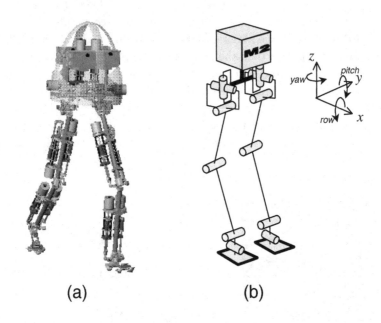

(a)   (b)

Fig. 6.2   Three dimensional biped: M2 (CAD drawing by Daniel Paluska).

In both systems, the joints are driven by force control actuators called Series Elastic Actuators [16]. They are assumed to be force sources in the simulation analysis. The model specifications for Spring Flamingo and M2 are given in Table 6.4 and 6.5, respectively, in Appendices 6.1 and 6.2. A dynamic simulation program called SD-FAST [17], which is based on the Newtonian mechanics for interconnected rigid bodies, is used for the dynamics simulation of these bipeds.

6.3   Control Architecture

A 3D bipedal walking system can be very complex to analyze from the perspectives of kinematics and dynamics. One way to overcome this problem is to partition the 3D motion space into a number of 2D motion spaces. One common approach is to project the walking motion onto the three orthogonal plane: sagittal, frontal and transverse. If we could assume that these projected motions can be independently controlled, we could decompose the overall walking control task into sagittal, frontal and transverse plane motion controls. This reduces significantly the complexity of the problem.

In this study, we propose a general control architecture for 3D dynamic bipedal walking which is based on a divide-and-conquer framework. We first decompose the overall walking control task into the motion controls in each of the orthogonal planes. To further reduce the complexity, we subdivide the motion control task in each plane into small subtasks which can be assigned to the stance leg or the swing leg accordingly. Learning tools are applied to these subtasks only when necessary.

This study demonstrates how the control architecture can be applied to the sagittal plane motion-control task. The sagittal plane motion-control task can be decomposed into the following subtasks: 1) to maintain body pitch; 2) to maintain body height; and 3) to maintain desired walking (average) speed. It is relatively easy to achieve the first and second subtasks in the sagittal plane. However, the third subtask is much more complex. It is directly associated with gait stabilization.

There are a number of speed control mechanisms in the sagittal plane. They can be partitioned into two main classes, viz, *global* and *local*. A *global* speed control mechanism can achieve speed control all by itself. However, a *local* speed control mechanism can only achieve limited speed control. Therefore, it cannot achieve the walking speed control all by itself.

Examples of *local* speed control mechanisms are those that are based on the double support phase [18], the biped's torso (using torso as a limited reaction wheel), the stance ankle, etc. In contrast, there is only one global speed control mechanism. It is based on swing leg control. If the swing leg control is badly executed, other (local) speed control mechanisms may not be capable of preventing the biped from falling. Swing leg control, therefore, is the key mechanism for walking speed control. The question is how to execute the swing leg in the sagittal plane such that stable walking can be achieved.

Although the swing leg motion is an important determinant for the stabilization of the walking speed, it is difficult to obtain a closed form expression for it. This is because the dynamic equations of a biped are complex and nonlinear. This is further complicated by the unpowered DOF between the stance foot and the ground during the single support phase [19]; and the discontinuities caused by the impact at the support exchange. The overall motion of the system is usually strongly coupled to swing leg motion unless the leg's inertia is negligible.

Some research assumes that the effect of the swing leg is negligible. Thus, the system can be represented by an inverted pendulum. However, this assumption becomes invalid if the mass of the leg is significant, for example, during fast walking. The swing leg dynamics may also be utilized for walking control, for example, to create reaction torque at the hip. If its effect is ignored, no such utilization is possible. The configuration of the swing leg also affects the projected center of gravity of the biped, thus changing the gravitational moment about the supporting leg. In view of the difficulty, we adopt a learning tool to assist in the swing leg motion planning.

6.4   Key Implementation Tools

This section describes the tools that are adopted for the control algorithm implementations. A control "language" for legged locomotion called Virtual Model Control (VMC) [18] is used for the motion control. This section also describes a reinforcement learning algorithm called Q-learning [20] which is adopted for the learning task.

6.4.1   *Virtual Model Control*

Virtual Model Control (VMC) [18] is a control language for legged locomotion. It allows us to work in a more intuitive space (e.g., Cartesian space) instead of a space (e.g. joint space or actuator space) that suits the robot. In the Virtual Model Control (VMC) approach, a control algorithm is constructed by applying virtual components (which usually have corresponding physical parts, for example, spring, damper, etc) at some strategic locations of the biped. The virtual components interact with the biped and generate a set of virtual forces based on their constitutive equations. Then these forces (represented by a vector $\vec{f}$) can be transformed into the joint space (represented by a vector $\vec{\tau}$) by a Jacobian matrix $_{B}^{A}J$:

$$\vec{\tau} = {_{B}^{A}}J^{T} {_{B}^{A}}\vec{f} \tag{6.1}$$

where $_{B}^{A}J$ is the Jacobian matrix which transforms the differential variation in the joint space into the differential variation of reference frame B with respect to reference frame A and the superscript ^{T} denotes the transpose of a matrix. In the Virtual Model Control approach, frame B and frame A

correspond to the action frame and reaction frame, respectively.

By changing the set points of the virtual components, an inequilibrium can be created in the system. The system will then adjust itself so that a new equilibrium position is reached. The biped will behave dynamically as if the actual physical components are attached to it provided that the actuators are perfect torque sources.

6.4.2 *Q-Learning*

Reinforcement learning is a class of learning problem in which a learner (agent) learns to achieve a goal through *trial-and-error* interactions with the environment. The learning agent learns only from reward information, and it is not told how to achieve the task. From failure experience, the learning agent reinforces its knowledge so that success can be attained in future trials.

Let's assume that the environment is stationary and the learning task can be modelled to be a fully observable Markov decision process (MDP) that has finite state and action sets. One popular algorithm for MDP that is the Q-learning algorithm by Watkins and Dayan [20]. It recursively estimates a scalar function called optimal Q-factors ($Q^*(i, u)$) from experience obtained at every stage, where i and u denote the state and corresponding action, respectively. The experience is in the form of immediate reward sequence, $r(i_t, u_t, i_{t+1})$ ($t = 0, 1, 2, \ldots$). $Q^*(i, u)$ gives the expected return when the agent takes the action u in the state i and adopts an optimal policy π^* thereafter. Based on $Q^*(i, u)$, an optimal policy π^* can easily be derived by simply taking any action u that maximizes $Q^*(i, u)$ over the action set $U(i)$.

For the discounted problem, the single-step sample update equation for $Q(i, u)$ is given as follows [20]:

$$
\begin{aligned}
Q_{t+1}(i_t, u_t) \leftarrow{} & Q_t(i_t, u_t) + \alpha_t(i_t, u_t)[r(i_t, u_t, i_{t+1}) \\
& + \gamma \max_{u \in U(i_{t+1})} Q_t(i_{t+1}, u) - Q_t(i_t, u_t)]
\end{aligned}
\tag{6.2}
$$

where the subscript indicates the stage number; $r(i_t, u_t, i_{t+1})$ denotes the immediate reward received due to the action u_t taken which causes the transition from state i_t to i_{t+1}; $\alpha \in [0, 1)$ denotes the step-size parameter for the update; $\gamma \in [0, 1)$ denotes the discount rate. Equation (6.2) updates $Q(i_t, u_t)$ based on the immediate reward $r(i_t, u_t, i_{t+1})$ and the maximum value of $Q(i_{t+1}, u)$ over all $u \in U(i_{t+1})$.

In most reinforcement learning implementation, there is an issue concerning the trade-off between "exploration" and "exploitation" [21]. It is the balance between trusting that the information obtained so far is sufficient to make the right decision (exploitation) and trying to get more information so that better decisions can be made (exploration). For example, when a robot faces an unknown environment, it has to first explore the environment to acquire some knowledge about the environment. The experience acquired must also be used (exploited) for action selection to maximize the rewards.

One method that allows a structured exploration is called ϵ-greedy method [21]. In this method, the agent selects an action by maximizing the current set of $Q(i, u)$ over the action set with probability equal to $(1 - \epsilon)$ (where ϵ is usually a small number). Otherwise, the agent randomly (uniformly) selects an action from the action set $U(i)$. A greedy policy is one in which $\epsilon = 0$.

In a reinforcement learning problem, it is required to identify those actions that contribute to a failure (success) and properly discredit (credit) them. This is called the credit assignment or delayed reward problem. If an immediate reward function that correctly evaluates the present action taken can be obtained, it is not necessary to consider any delayed reward. If the reward function is not precise or if the action has an effect that lasts more than one stage, it is required to correctly assign credit to it based on the subsequent rewards received.

In the Q-learning algorithm, the discount factor γ determines the credit assignment structure. When γ is zero, only immediate reward (myopic) is considered. That is, any rewards generated by subsequent states will not have any influence on the return computed at the present state. Usually, if a reinforcement learning problem can be posed such that $\gamma = 0$ works fine, it is much simpler and can be solved more quickly. When γ is large (close to one), the future rewards have significant contribution to the return computation for the present action. The agent is said to be farsighted since it looks far ahead of time when evaluating a state-action pair.

6.4.3   *Q-Learning Algorithm Using Function Approximator for Q-Factors*

This subsection describes the Q-learning algorithm that uses a function approximator to store the Q-factors. The function approximator (e.g., multi-layered perceptron, radial basis functions, CMAC etc.) is used as a more compact but approximate representation of $Q(i, u)$. The purposes of using a function approximator are mainly to reduce memory requirement and to enable generalization [21], [22]. These are desirable especially for high-dimensional and continuous state-action space.

In this chapter, Cerebellar Model Articulation Controller (CMAC) [23] is used as a multivariable function approximator for the Q-factors in the Q-learning algorithms. CMAC has the advantage of having not only local generalization, but also being low in computation. The Q-learning algorithm using CMAC as the Q-factor function approximator is summarized as in Figure 6.3 [20], [21]. There are many ways to implement a CMAC network. In this study, the original Albus's CMAC implementation is adopted and the corresponding algorithm is well described in [24].

INITIALIZE weights of CMAC
REPEAT (for each trial):
        Initialize i
        REPEAT (for each step in the trial)
                Select action u under state i using policy
                (say ϵ-greedy) based on $\hat{Q}$
                Take action u,
                Detect new state i' and reward r
                IF i' not a failure state
                    $\delta \leftarrow r + \gamma \max_{u'} \hat{Q}(i', u') - \hat{Q}(i, u)$
                    Update weights of CMAC based on δ
                    $i \leftarrow i'$;
                ELSE ($r = r_f$)
                    $\delta \leftarrow r_f - Q(\hat{i}, u)$
                    Update weights of CMAC based on δ
             UNTIL failure encountered or target achieved
    UNTIL target achieved or number of trials exceed a preset limit

---

Fig. 6.3   Q-learning algorithm using CMAC to represent Q-factors.

6.5   Implementations

This section describes the control algorithm for the sagittal plane motion of the bipeds. It is divided into the following subtasks: 1) the height control of the body; 2) the pitch control of the body; and 3) the forward velocity control of the body. For M2, those joints whose axes are perpendicular to the sagittal plane are utilized for these subtasks.

Consider the "bird-like" and "bent-leg" walking posture and assume that the desired walking height and body pitch angle are constant. The height control and body posture control subtasks can be achieved using the Virtual Model Control without learning. For the horizontal speed control, it is mainly determined by the swing leg control. Since there is no analytical solution for the desired swing leg behavior, the Q-learning algorithm is chosen to learn the key parameter(s) for the swing leg control.

The following subsection describes the implementation of the control algorithm for the stance leg and the swing leg using the Virtual Model Control approach. The subsequent subsection describes the application of the Q-learning algorithm to learn the key swing leg parameters for the sagittal plane motion so that stable walking cycle can be achieved.

6.5.1   *Virtual Model Control Implementation*

In the sagittal plane motion control, the stance leg's joints are used to achieve the height control and body pitch control of the biped during walking. The swing leg's joints are used to execute the swing leg strategy whose parameters are obtained by learning. The control algorithm for the stance leg and the swing leg can be considered separately as follows.

The stance leg algorithm is similar to the single-support algorithm in [18]. The height control is achieved by a virtual spring-damper attached vertically between the hip and the ground. It generates a virtual vertical force f_z at the hip. For the body pitch control, a virtual rotational spring-damper can be applied at the hip. This generates a virtual moment m_α about the hip. The parameters of these virtual components (as summarized in Table 6.1)are tuned by trial-and-error approach. The virtual forces (f_z and m_α) are transformed into the desired torques for the stance-knee joint and the stance-hip pitch-joint (τ_k and τ_{hp}, respectively) using a transformation matrix [18].

There are many ways to implement the swing leg strategy. This study

Table 6.1 Parameters of the virtual components for the stance leg (in the sagittal plane).

Parameter	Notation
1. Spring stiffness (in z direction)	k_z
2. Damping coefficient (in z direction)	b_z
3. Spring stiffness (in α direction)	k_α
4. Damping coefficient (in α direction)	b_α

adopts one strategy that involves specifying a sequence of swing foot trajectory in the Cartesian space as follows:

(1) Lift the swing foot to a specific lift height, l_{lh}, from 0 to t_1 seconds. A third degree polynomial is used for this trajectory planning.
(2) Swing the leg forward to a specific horizontal position once the foot is in the air. Again, a third degree polynomial is used for this trajectory planning.
(3) After T_{swing} seconds, the swing foot starts to descend to the ground while maintaining its desired horizontal position.

Given the sequence of the swing foot trajectories, two sets of linear spring-damper virtual components which are along the X-axis (horizontal) and the Z-axis (vertical) of the inertia frame respectively can be adopted. Each of them is attached between the ankle of the swing leg and the trajectory path. The virtual forces in these virtual components are then transformed by a Jacobian equation [18] to the respective joint torques of the swing leg.

The swing leg strategy adopted has several parameters that need to be set. They are lift height, swing time and end position of the swing foot. The gait stability of the dynamic walking depends on the set of parameter values chosen. In the next subsection, a reinforcement learning algorithm will be used to determine the values of selected parameters so that a walking goal can be achieved.

6.5.2 *Reinforcement Learning to Learn Key Swing Leg's Parameter*

The key parameters for the swing leg strategy introduced in Subsection 6.5.1 are the lift height, swing time and the end position of the swing foot. In this investigation, the lift height is set to be constant throughout the walking to simplify the problem. That is, the only key parameters considered are the swing time and the end position of the swing foot. There are three possible learning implementations. In the first implementation, the swing time is fixed and the learning agent learns the end position of the swing foot. In the second implementation, the end position of the swing leg is fixed and the learning agent learns the swing time. In the third implementation, a learning agent learns the right combinations of these parameters.

In this investigation, the first two implementations are considered. The choice between the first and the second implementations depends on the terrain type. For example, if the actual foot placement or step length is not important (say in the case of the level ground walking), it is possible to simply set the swing time and learn the end position of the swing foot with reference to the body. However, if the actual foot placement is important (say in the case of the walking on stairs), it is better to specify the desired step length and learn the swing time that results in stable gait.

This subsection presents the approach whereby the decision for the swing leg parameter can be posed as a discrete-time delayed reward (or punishment) problem. By trial-and-error, the biped is supposed to learn the right swing leg parameter at each step so that stable walking is resulted.

First, We need to identify the state variables for the learning task. A reward function is also set up for the learning. After that, an appropriate reinforcement learning algorithm has to be chosen. The following subsubsections are devoted to these tasks.

6.5.2.1 *State variables*

The hip height and the body pitch are assumed to be constant during the walking and hence they can be excluded from the state variables set in the learning implementation. For constant desired walking speed (average), the following variables are identified to be the state variables for the learning implementation (in the sagittal plane):

- Velocity of the hip in the x-direction, $\dot{x}^+$;
- x-coordinate of the previous swing ankle measured with reference to the hip, x_{ha}^+;
- Step length, l_{step}^+.

Superscript $+$ indicates that the state variable is measured or computed at the beginning of a new single support phase.

The choice of the first two state variables are obvious. They represent the generalized coordinate for the body, assuming the height and the body pitch are constant. The third state variable (the step length, l_{step}^+) is required to take the dynamics of the swing leg into account. If the swing leg dynamics were negligible, this state variable will be ignorable.

6.5.2.2 *Reward function and reinforcement learning algorithm*

The biped aims to select a good value for the swing leg parameter for each consecutive step so that it achieves stable walking. A reward function that correctly defines this objective is critical in the reinforcement learning algorithm. If it is not correctly defined, the learning task may not converge.

Let's assume that there are minimum heuristics available for the implementation. A simple reward function can be formulated by defining a failure state to be one whose walking speed is above an upper bound V_u or below a lower bound V_l; or the hip height z is lower than a lower bound Z_l:

$$r = \begin{cases} 0 & \text{for } V_l \leq \dot{x} \leq V_u \text{ and } z \geq Z_l \\ R_f & \text{otherwise (failure).} \end{cases} \qquad (6.3)$$

where R_f is a negative constant corresponding to punishment. There is no immediate reward or punishment unless a failure state is encountered. This is a *delayed reward* problem and the agent has to learn to avoid the failure states and to avoid those states that inevitably lead to the failure states.

The Q-learning algorithm that uses CMAC network as the function approximator is adopted for the implementations. The learning target is for the biped to achieve say 100 seconds of walking without encountering the failure states. Although the action sets for this application can be continuous, they are discretized so that the "maximum" Q-factor can be conveniently searched. The discretization resolution depends on the computational capability of the microprocessor and the nature of the problem. For example, in the case of the foot placement selection (see Subsection 6.6.1), the discretization resolution is chosen to be $0.01m$ that is fine enough so as it does not affect the walking behavior nor requires high computation.

The mechanism of the learning can be briefly summarized as follows. Before learning begins, all the Q-factors are initialized to an optimistic value, say zero (by setting all the weights of the CMAC network to zero). Each learning iteration is from an initial state to either a failure state or any state in which the learning target is achieved. At the beginning of the first learning iteration, all the actions in the discrete action set will be equally good. In the event of a tie, the first action encountered in the search algorithm that has the maximum Q-factor will be adopted. Whenever a failure state is encountered, a new learning iteration is initialized. The Q-factor for the state-action pair that leads to the failure state will be downgraded. If all the actions at a given state lead to a failure state, the learning agent will avoid a state-action pair that leads to that state. As such, the learning agent gradually identifies those state-action pairs that lead to a bad outcome.

6.6   Simulation Studies and Discussion

This section presents two simulation studies. The first study illustrates the positive effect of a local speed control mechanism in the sagittal plane algorithm. The learning rate in the reinforcement learning algorithm will be shown to be greatly increased when the control algorithm includes such a mechanism.

The second study illustrates the generality of the proposed algorithm. We apply the sagittal plane algorithm to the two different bipeds introduced in Section 6.2. The simulation result will show that both bipeds, even though they have different length and inertia parameters, can learn equally well to walk.

6.6.1   *Effect of Local Speed Control on Learning Rate*

In Section 6.3, the speed control mechanisms in the sagittal plane were divided into *local* and *global*. Although the local control mechanisms are not able to achieve the speed control by themselves, appropriate use of them may help to improve the performance of the control algorithm. In this subsection, the local speed control mechanism which is based on the stance ankle (Figure 6.4) will be demonstrated to significantly reduce the learning time.

The control law applied to the stance-ankle pitch joint to generate the required torque τ_a is as follows:

$$\tau_a = B_a(\dot{x} - \dot{x}^d) \tag{6.4}$$

where $\dot{x}$ is the velocity of the hip in the x-direction, B_a is a constant gain and superscript d indicates the desired value. The torque τ_a is bounded

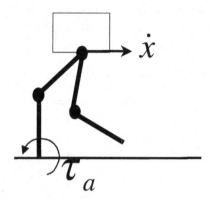

Fig. 6.4   A local speed control mechanism based on stance ankle joint.

within upper and lower limits to prevent the stance leg's foot from tipping at the toe or heel. In the algorithm, a simple static analysis is used to generate the bounds:

$$\tau_{au} = K_{au}Mgl_{at\_x} \tag{6.5}$$

$$\tau_{al} = -K_{al}Mgl_{ah\_x} \tag{6.6}$$

where τ_{au} and τ_{al} are the upper and lower bounds of τ_a, respectively; M is the total mass of the biped; g is the gravitational constant; l_{at_x} is the horizontal distance between the ankle and the front end of the foot; l_{ah_x} is the horizontal distance between the ankle and the rear end of the foot; K_{au} and K_{al} are positive discount factors to modulate the values of the bounds.

To test the effectiveness of the local control, a control algorithm was constructed as described in the previous section. It was applied to the simulated Spring Flamingo. The swing time T_s was fixed at $0.3second$. The Q-learning algorithm was used to learn the desired end position of the swing foot. In this particular implementation, the desired position of the swing foot was measured horizontally from the hip. It was selected from a discrete action set U ($= \{0.01n|$ for $0 \le 0.01n \le 0.2$ and $n \in \mathbf{Z}\}$).

The values of V_u, V_l and Z_l used in the simple reward function (Equation 6.3) to determine the failure conditions were mainly chosen by intuition. These parameter values (including the upper and lower bounds for the action set) were tuned by trial-and-error. The performance of the algorithm was not very sensitive to these values. For the simple reward function, a farsighted learning agent was used. The discount factor γ was set to a value close to one (say 0.9). The details of this implementation are summarized in Table 6.2.

Table 6.2   Reinforcement learning implementation for the sagittal plane: S_1.

Description	Remark
Implementation code	S_1
Swing time, T_s	constant
Desired walking speed	constant
Learning output (action)	Horizontal end position of the swing foot with reference to the hip, x^f_{ha}
Learning target	100s of walking
Key parameters:	
Reward function	Equation 6.3
$\quad$ Upper bound for the hip velocity V_u	$-0.1m/s$
$\quad$ Lower bound for the hip velocity V_l	$1.3m/s$
$\quad$ Lower bound for the hip height Z_l	0.5
$\quad R_f$	-1
Discount factor γ	0.9 (farsighted)
Action set U	$\{0.01n|$ for $0 \le 0.01n \le 0.2$ and $n \in \mathbf{Z}\}$
Policy	Greedy

The starting posture (standing posture) and initial walking speed ($0.4m/s$) were the same for every iteration during the learning process. The desired walking speed was $0.7m/s$. Two learning curves are shown in Figure 6.5. The learning curve in dashdot-line corresponds to the case where the local control was added. It achieved 100 seconds of walking at $16th$ iterations. The learning curve in solid-line corresponds to the simulation result in which the local control was not added (the stance ankle was limp). From the learning curves, it can be deduced that proper application of the stance ankle torque can speed up the learning rate for the walking.

The graphs in Figure 6.6 show the action (x^f_{ha}) sequences generated by the learning agent (with local control at the stance ankle) for the 1st, 5th, 10th, 15th and 16th learning iterations. The corresponding forward velocity ($\dot{x}$) profiles are also included in the same figure. Except for the 16th iteration in which the learning target was achieved, all other iterations prior to the 16th iteration were terminated due to failure encounter. At the 16th iteration, the first portions (before six seconds) of the graphs for both $\dot{x}$ and x^f_{ha} were rather chaotic. After that, the motion converged to a steady cycle.

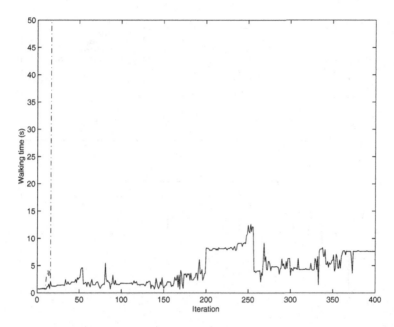

Fig. 6.5 The graphs show that the local control based on the stance ankle increases the learning rate. The dashdot-line graph corresponds to the learning curve with the local control at the ankle. The solid-line graph is without the local control. In both cases, the starting posture (standing posture) and the initial speed ($0.4m/s$) are the same for each training iteration.

To verify that the local control at the stance ankle pitch joint did consistently result in a good learning rate, the same implementation that used the local control was applied to the simulated biped for different swing times (0.2, 0.3, 0.4 and 0.5 second). The initial walking velocity was set to zero. All other parameters were set as before. Figure 6.7 shows the learning curves for each of the swing times. In the worst case, the biped could achieve 100 seconds of continuous walking within 90 iterations.

In summary, the proper usage of the local speed control mechanism based on the stance ankle can help to increase the learning rate. It may be partly because the mechanism has increased the space of successful solutions. As such, the learning algorithm can find a solution much faster. The mechanism can also be viewed to possess an "entrainment" property. This property permits the learning agent to simply generate a coarse foot placement which results in a walking speed that is sufficiently close to the desired value. Then, the mechanism try to modulate the walking speed so

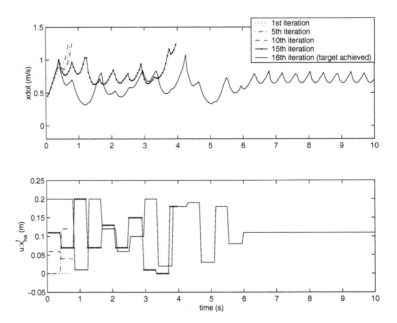

Fig. 6.6   The graphs show the action (x_{ha}^f) sequences generated by the learning agent (with local control at the stance ankle) for the 1st, 5th, 10th, 15th and 16th learning iterations. The corresponding forward velocity $(\dot{x})$ profiles are also included. The learning target is achieved at the 16th iteration. The starting posture (standing posture) and the initial speed $(0.4m/s)$ are the same for each training iteration.

that the speed is even closer to the desired value. It is interesting to note that the biped was able to achieve the desired walking speed quite well even though the reward function did not take it into account. All the sagittal plane implementations in the subsequence subsection will have such a local speed control mechanism.

6.6.2   *Generality of Proposed Algorithm*

This subsection demonstrates the generality of the sagittal plane control algorithm in terms of its applicability across bipeds of different mass and length parameters. The desired end position of the swing foot was specified based on the desired step length. Such an implementation is needed when the biped is required to select its foot placement according to constraints due to the terrain. One example is when the biped is required to walk on a stair.

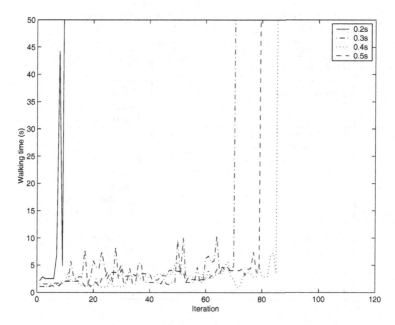

Fig. 6.7 The learning curves of the simulated Spring Flamingo correspond to different swing times (0.2, 0.3, 0.4 and 0.5 second). The local control based on the stance ankle is used. The starting posture (standing posture) and the initial speed ($0m/s$) for each training iteration are the same for all the cases.

Given a desired step length, the biped was required to learn an appropriate swing time, T_s. For a given step length, different swing times result in different postures, and hence the overall gait stability. A control algorithm very similar to that in the previous subsection was constructed for this analysis. The difference was mainly in the reinforcement learning implementation. Here, the step length was given and the learning agent learned the swing time. The learning agent was farsighted as in the previous implementation. The key parameters of the reinforcement learning implementation is summarized in Table III.

The control algorithm was applied to both the simulated Spring Flamingo (SF) and M2 (constrained to the sagittal plane). The desired step length was constant ($0.3m$). The starting posture (standing position) and the initial hip velocity ($0.4m/s$) were the same for each trial. The desired hip velocities were $0.7m/s$ for both bipeds. The desired heights were $0.74m$ for Spring Flamingo and $0.84m$ for M2. Again, the local speed control based on the stance ankle was adopted to assist in the velocity control.

Table 6.3   Reinforcement learning implementation for the sagittal plane: S_2.

Description	Remark
Implementation code	S_2
Step length	constant
Desired walking speed	constant
Learning output (action)	Swing time T_s
Learning target	$100s$ of walking
Key parameters for RL algorithm:	
Reward function	Equation 6.3
V_u	$-0.1m/s$
V_l	$1.5m/s$
Z_l	$0.5m$ (SF)$/0.5m$ (M2)
R_f	-1
Discount factor γ	0.9 (farsighted)
Action set U	$\{0.02n \vert$ for $0.2 \leq 0.02n \leq 1$ and $n \in \mathbf{Z}\}$
Policy	Greedy

The results are shown in Figure 6.8. The learning speeds were comparable, and this demonstrated the generality of the proposed algorithm in terms of its application to bipeds of different inertia and length parameters. The results also demonstrated that the bipeds could achieve stable walking even though the step length was constrained.

A stick diagram of the dynamic walking of Spring Flamingo is plotted in Figure 6.9. The simulation data for the Spring Flamingo implementation (after the learning target has been achieved) is shown in Figure 6.10. Those variables with "i" preceding them were the state variables for the reinforcement learning algorithm and that with "u" preceding it was the action variable. The top graph shows the states of the state machine (state 5 and state 6 corresponded to the right support and the left support, respectively; state 7 corresponded to the right-to-left transition; and state 8 for the left-to-right transition). The horizontal dashed lines in the second and third graphs indicated the desired values.

From the second and third graphs, it is observed that the actual hip height and the forward velocity were well behaved, and they were close to the desired value. Especially for the forward velocity, although the reward function did not take into account the desired forward velocity, its average value was very close to the desired value. This outcome could be attributed to the local control at the stance ankle as discussed in Subsection 6.6.1.

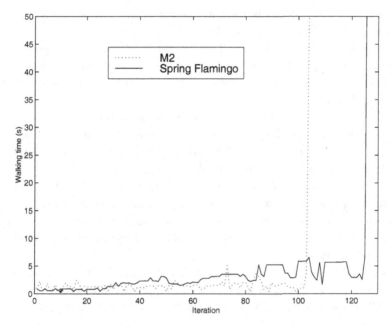

Fig. 6.8 Learning curves for the simulated M2 (constrained to the sagittal plane) and Spring Flamingo (using implementation S_2).

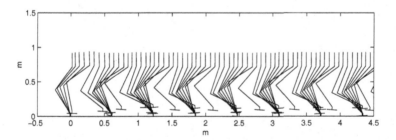

Fig. 6.9 Stick diagram of the dynamic walking of the simulated Spring Flamingo after the learning target was achieved (using implementation S_2). Only the left leg is shown in the diagram, and the images are 0.1 second apart.

6.7 Summary

This chapter first presented the framework of a control architecture that was based on a divide-and-conquer approach. It illustrated how the architecture could be applied to generate the sagittal-plane motion-control algorithm.

It also presented the tools used in the algorithm. In particular, the Q-learning algorithm was applied to learn the parameter of the swing leg strategy in the sagittal plane. The Virtual Model Control approach was used to generate the desired torques for all the joints in the same plane. The resulting sagittal plane motion control algorithm was applied to the two simulated bipedal robots.

The simulation studies illustrated the usage of the local speed control mechanism based on the stance ankle to significantly improve the learning rate for the implementation. They illustrated the importance of the stance ankle even though it was not capable of achieving global speed control. Next, the generality of the algorithm was demonstrated. The algorithm could be applied without major retunings across different bipeds that had the same structure but different inertia and length parameters. Each biped simply went through its learning process to achieve the walking motion in the sagittal plane.

In this chapter, we have presented the sagittal-plane motion-control implementation. The frontal and transverse plane motion-control, and the overall three-dimensional walking algorithm can be found in [25].

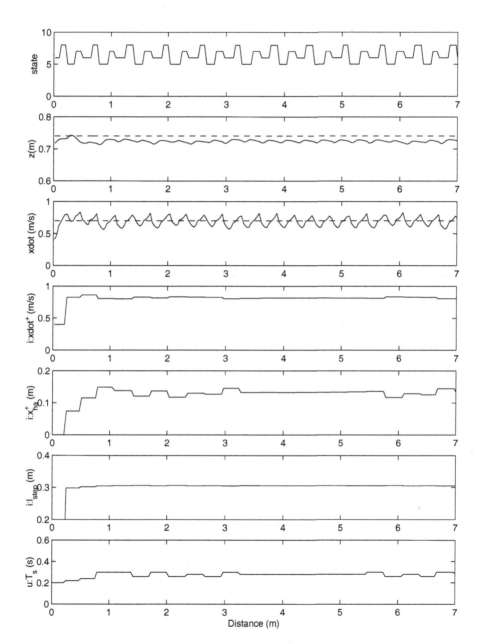

Fig. 6.10 The simulation data for the dynamic walking of the simulated Spring Flamingo after the learning target was achieved (using implementation S_2).

Bibliography

[1] F. Gubina and H. Hemami R. B. McGhee, On the dynamic stability of biped locomotion, *IEEE Transactions on Biomedical Engineering*, vol. BME-21, no. 2, pp. 102–108, Mar 1974.

[2] S. Arimoto and F. Miyazaki, Biped locomotion robots, *Japan Annual Review in Electronics, Computers and Telecommunications*, vol. 12, pp. 194–205, 1984.

[3] F. Miyazaki and S. Arimoto, A control theoretic study on dynamical biped locomotion, *ASME Journal of Dynamic Systems, Measurement, and Control*, vol. 102, pp. 233–239, 1980.

[4] H. Miura and I. Shimoyama, Dynamic walk of a biped, *International Journal of Robotics Research*, vol. 3, no. 2, pp. 60–74, 1984.

[5] J. S. Bay and H. Hemami, Modeling of a neural pattern generator with coupled nonlinear oscillators, *IEEE Transactions on Biomedical Engineering*, vol. BME-34, no. 4, pp. 297–306, April 1987.

[6] K. Hirai, M. Hirose, Y. Haikawa, and T. Takenaka, The development of honda humanoid robot, *IEEE International Conference on Robotics and Automation*, pp. 1321–1326, 1998.

[7] Y. Hurmuzlu, Dynamics of bipedal gait: Part i - objective functions and the contact event of a planar ve-link biped, *Journal of Applied Mechanics*, vol. 60, pp. 331–336, 1993.

[8] S. Kajita, T. Yamura, and A. Kobayashi, Dynamic walking control of a biped robot along a potential energy conserving orbit, *IEEE Transactions on Robotics and Automation*, vol. 6, no. 1, pp. 431–438, 1992.

[9] J. A. Golden and Y. F. Zheng, Gait synthesis for the sd-2 biped robot to climb stairs, *International Journal of Robotics and Automation*, vol. 5, no. 4, pp. 149–159, 1990.

[10] G. A. Pratt, Legged robots at mit - what's new since raibert, *2nd International Conference on Climbing and Walking Robots*, pp. 29–33, 1999.

[11] S. Kawamura, T. Kawamura, D. Fujino, and S. Arimoto, Realization of biped locomotion by motion pattern learning, *Journal of the Robotics Society of Japan*, vol. 3, no. 3, pp. 177–187, 1985.

[12] J. Yamaguchi, A. Takanishi, and I. Kato, Development of a biped walking

robot compensating for three-axis moment by trunk motion, *IEEE International Conference on Intelligent Robots and Systems*, pp. 561–566, 1993.

[13] H. Wang, T. T. Lee, and W. A. Grover, A neuromorphic controller for a three-link biped robot, *IEEE Transactions on Systems, Man, and Cybernetics*, vol. 22, no. 1, pp. 164–169, Jan/Feb 1992.

[14] T. W. Miller, Real time neural network control of a biped walking robot, *IEEE Control Systems Magazine*, pp. Feb:41–48, 1994.

[15] H. Benbrahim and J. A. Franklin, Biped dynamic walking using reinforcement learning, *Robotics and Autonomous Systems*, vol. 22, pp. 283–302, 1997.

[16] G. A. Pratt and M. M. Williamson, Series elastic actuators, *IEEE International Conference on Intelligent Robots and Systems*, vol. 1, pp. 399–406, 1995.

[17] D. E. Rosenthal and M. A. Sherman, High performance multibody simulations via symbolic equation manipulation and kane's method., *Journal of Astronautical Sciences*, vol. 34, no. 3, pp. 223–239, 1986.

[18] J. Pratt, C.-M. Chew, A. Torres, P. Dilworth, and G. Pratt, Virtual model control: An intuitive approach for bipedal locomotion, *International Journal of Robotics Research*, vol. 20, no. 2, pp. 129–143, 2001.

[19] M. Vukobratovic, B. Borovac, D. Surla, and D. Stokic, *Biped Locomotion: Dynamics, Stability, Control, and Applications*, Springer-Verlag, Berlin, 1990.

[20] C. J. C. H. Watkins and P. Dayan, Q-learning, *Machine Learning*, vol. 8, pp. 279–292, 1992.

[21] R. S. Sutton and A. G. Barto, *Reinforcement Learning: An Introduction*, MIT Press, Cambridge, MA, 1998.

[22] D. Bertsekas and J. Tsitsiklis, *Neuro-Dynamic Programming*, Athena Scientific, Belmont, MA, 1996.

[23] J. S. Albus, *Brain, Behavior and Robotics*, chapter 6, pp. 139–179, BYTE Books. McGraw-Hill, Peterborough, NH, 1981.

[24] T. W. Miller and F. H. Glanz, The university of new hampshire implementation of the cerebellar model arithmetic computer - cmac, *Unpublished reference guide to a CMAC program*, August 1994.

[25] C.-M. Chew and G. A. Pratt, Frontal plane algorithms for dynamic bipedal walking, *Robotica*, vol. 22, pp. 29–39, 2004.

[26] C.-M. Chew and G. A. Pratt, Dynamic bipedal walking assisted by learning, *Robotica*, vol. 20, pp. 477–491, 2002.

[27] C.-M. Chew and G. A. Pratt, Adaptation to load variations of a planar biped: Height control using robust adaptive control, *Robotics and Autonomous Systems*, vol. 35, no. 1, pp. 1–22, 2001.

Appendix 6.1: Specifications for Spring Flamingo

Table 6.4   Specifications of Spring Flamingo.

Description	Value
Total Mass	12.04kg
Body Mass	10.00kg
Thigh Mass	0.46kg
Shin Mass	0.31kg
Foot Mass	0.25kg
Body's Moment of Inertia about COM	0.100kgm^2
Thigh's Moment of Inertia about COM	0.0125kgm^2
Shin's Moment of Inertia about COM	0.00949kgm^2
Foot's Moment of Inertia about COM	0.00134kgm^2
Thigh Length	0.42m
Shin Length	0.42m
Ankle Height	0.04m
Foot Length	0.15m

Appendix 6.2: Specifications for M2

Table 6.5   Specifications of M2.

Description	Value
Total Mass	25.00kg
Body Mass	12.82kg
Thigh Mass	2.74kg
Shin Mass	2.69kg
Foot Mass	0.66kg
Body's Principal Moment of Inertia	
x-axis	0.230kgm^2
y-axis	0.230kgm^2
z-axis	0.206kgm^2
Thigh's Principal Moment of Inertia	
x-axis	0.0443kgm^2
y-axis	0.0443kgm^2
z-axis	0.00356kgm^2
Shin's Principal Moment of Inertia	
x-axis	0.0542kgm^2
y-axis	0.0542kgm^2
z-axis	0.00346kgm^2
Foot's Principal Moment of Inertia	
x-axis	0.000363kgm^2
y-axis	0.00152kgm^2
z-axis	0.00170kgm^2
Hip Spacing	0.184m
Thigh Length	0.432m
Shin Length	0.432m
Ankle Height (from ankle's pitch axis)	0.0764m
Foot Length	0.20m
Foot Width	0.089m

Chapter 7

Swing Time Generation for Bipedal Walking Control Using GA tuned Fuzzy Logic Controller

Dynamic bipedal walking control has long been a key focus in humanoid research. As bipedal walking behavior may involve some intuitions, it is possible to seek the strength of fuzzy logic (FL) which is a knowledge-based control methodology that mimics how a person would make decisions based on imprecise information. In this work, a Genetic algorithm (GA) tuned Fuzzy Logic Controller (FLC) is proposed for bipedal walking control implementation. The basic structure of FLC is constructed based on the linear inverted pendulum model. GA is implemented to search and optimize the FLC parameters. Once optimized, the FLC outputs a sequence of swing times which results in continuous stable walking.

7.1 Introduction

Many algorithms have been proposed for bipedal walking task. Numerous researchers investigated biped dynamics to reduce the complexity and proposed simplified models ([5],[8],[9]). Bipedal walking algorithms can be categorized into static and dynamic types. Stability is a basic requirement of these algorithms. For static walking, stable behavior is realized by ensuring that the projection of the center-of-mass of the robot is always kept within the support area [12]. For dynamic walking, the approach is usually to generate a set of joints' trajectories that will result in an indicator, called the zero moment point (ZMP), to be within the support area [14]. The ZMP can be viewed to be the point on the ground through which the reaction force vector acts on the robot such that the coupled moment has zero horizontal components. Dynamic walking algorithms are preferable as it can achieve fast walking behaviors. However, it may not be easy to generate feasible set of joints' trajectory that may results in stable walking (one that results in the ZMP always within the support area.

Another approach for dynamic walking control is to make use of physical model of the robot. As most bipedal robot models are nonlinear, they usually do not have analytical solutions. Furthermore, their dynamics are often difficult to model accurately. Kajita et al. [5] have proposed a linear inverted pendulum model for bipedal walking control. This model has analytical solution which can be used to generate key walking parameters. However, it may not be generally applicable to physical robots, especially if the dynamics of the robots deviate significantly from this ideal model. This has prompted our research work to develop walking control algorithm for bipedal robots that can be generally applied to physical robots without the need to obtain an accurate models. Computational intelligence tools are adopted for such algorithms. Some form of learning is required to cater to the uncertainties and to generate behaviors that results in proper walking execution. There are many ways to implement a learning-based algorithm for bipedal walking and several interesting works has been proposed [2], [7].

As human walking behaviors have certain intuitions (for example, if one wants to walk fast, he may swing his leg faster), we may be able to exploit the strength of Fuzzy Logic Control (FLC). FLCs are knowledge-based systems that make use of fuzzy rules and fuzzy Membership Functions (MF) to incorporate the human knowledge into their systems. The definition of fuzzy rules and MF is affected by subjective decisions which have a great influence over the whole FLC performance. Some efforts have been made to improve system performance by incorporating evolutionary mechanisms to modify rules and/or MF. For example, the application of genetic algorithm (GA) to FLCs has recently produced some interesting works [1],[11],[10],[3]. Inspired by those innovative applications, this chapter investigates a learning control algorithm, which synthesizes fuzzy logic and genetic algorithm, to achieve continuous stable dynamic walking for a simulated planar bipedal robot which has similar mass distribution and geometric structure as a physical seven-link planar biped called Spring Flamingo (see Figure 7.1).

The rest of the chapter is organized as follows. Section 7.2 gives a brief introduction to FLC, GA and GA tuned FLC. A basic mathematical model for the bipedal walking is given in Section 7.3. Section 7.4 describes the proposed control architecture. Simulation results are presented in Section 7.5. Last section concludes the work and suggests future work.

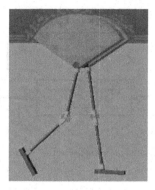

Fig. 7.1   7-link planar bipedal robot - Spring Flamingo.

7.2   Fuzzy Logic Control and Genetic Algorithm

7.2.1   *Fuzzy Logic Control (FLC)*

The concept of Fuzzy Logic (FL) was conceived by Lotfi Zadeh [15] as a way of processing data by allowing partial set membership rather than crisp set membership or non-membership. Zadeh reasoned that people do not require precise, numerical information input, and yet they are capable of achieve highly adaptive control. If feedback controllers could be programmed to accept noisy and imprecise inputs, they would be much more effective and perhaps easier to implement.

In this context, FL is a problem-solving methodology that provides a simple way to arrive at a definite conclusion based upon vague, ambiguous, imprecise, noisy or missing input information. FL's approach to control problems mimics how a person would make decisions to solve the problems.

FL incorporates a simple, rule-based approach (e.g. "IF X AND Y THEN Z") to solve a problem rather than attempting to model a system mathematically. The FL model is empirically-based, relying on an operator's experience rather than their technical understanding of the system.

A fuzzy controller consists of the fuzzification, defuzzification, rule-base and fuzzy inference engine as shown in Figure 7.2. The way the input and output variables are partitioned using fuzzy sets is called fuzzification. The relationship linking the input variable fuzzy sets and output variable fuzzy sets is called fuzzy rules. Procedures must be developed to combine these rules and to generate a real number, which can be used to control the plant. The combination of these rules is called fuzzy inference, and the procedure

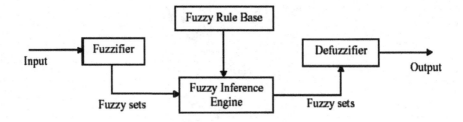

Fig. 7.2   Basic configuration of a fuzzy logic system.

to generate a real number is called defuzzification. Designing of a fuzzy logic controller typical includes the following steps:

- Identify the variables, i.e. inputs, states and outputs, of the plant.
- Partition the universe of discourse into a number of fuzzy subsets, assigning each a linguistic label. Assign or determine a membership function for each fuzzy subset.
- Assign the fuzzy relationships between the inputs' fuzzy subsets and the outputs fuzzy subsets, thus forming the rule-base.
- Normalize the domains.
- Fuzzification: To take the current normalized inputs and determine the degree to which they belong to each of the appropriate fuzzy sets via membership functions.
- Fuzzy inference: Use fuzzy approximate reasoning to infer the output contributed from each rule. Aggregate the fuzzy outputs recommended by each rule.
- Defuzzification: To obtain a scalar value of output from the resulted fuzzy set from fuzzy inference.
- Denormalization: To obtain the physical value of the control action.
- Test the system, evaluate the results, tune the rules and membership functions, and retest until satisfactory results are obtained.

7.2.2   *Genetic Algorithms (GAs)*

Genetic Algorithms (GAs) [4] are probabilistic search and optimization procedures based on natural genetics, working with finite strings (often referred as chromosomes) that represent the set of parameters of the problem. A fitness function is used to evaluate each of these strings. GAs are usually characterized by having:

- A coding scheme in the form of string or chromosome for each possible solution of the problem. .Each chromosome may consist of several genes which code various parameters of the problem.
- An initial set of solutions to the problem (initial population).
- An evaluation function that estimates the quality of each solution or string that composes the set of solutions (or population).
- A set of genetic operators that uses the information contained in a certain population (referred to as a generation) and a set of genetic parameters to create a new population (next generation).
- A termination condition to define the end of the genetic process.

The three basic genetic operators are reproduction, crossover and mutation. The reproduction operator creates a mating pool where strings are copied (or reproduced) from the old generation to await the crossover and mutation operations. Those strings with higher fitness values will have more copies in the mating pool. The crossover operator provides a mechanism for strings to exchange their components through a random process. This operator is applied to each pair of strings from the mating pool. The process consists of three steps: a pair of strings is randomly selected from the mating pool, a position along the string is selected uniformly at random, the elements following the crossing site of the strings are swapped between both strings. Mutation is the occasional alteration of a gene at a certain string position selected according to a mutation probability.

7.2.3   GA Tuned FLC

By means of GA, the performance of predefined FLCs can be improved through evolutionary optimization. Typical architecture for a GA tuned FLC is shown in Figure 7.3 [6], [10], [13].

When designing a FLC, the generations of the membership functions (MF) for each fuzzy set and the fuzzy rules are important. For MF, triangular or trapezoidal functions are usually employed. These functions are parameterized with several coefficients which will constitute the genes of the chromosome in the GA. This gene may be binary coded or real number.

GA is also used to modify the decision table of a FLC. A chromosome is formed from the decision table by going row-wise and coding each output fuzzy set as an integers, $0, 1, \ldots, n$, where n is the number of MF defined for the output variable of the FLC.

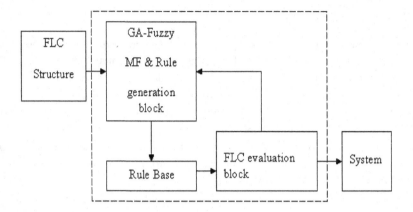

Fig. 7.3   Architecture of FLC with GA learning.

7.3   Linear Inverted Pendulum Model

To appreciate the bipedal walking dynamics, we study a simple mathematical model called the linear inverted pendulum proposed by [5]. It will also be used to examine the relationship among several key parameters for bipedal walking.

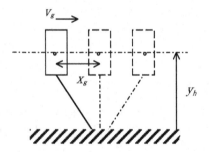

Fig. 7.4   Linear inverted pendulum model.

In this model, the body is approximated by a point mass. The legs are assumed to be massless. The body is assumed to move along a linear path parallel to the ground (Figure 7.4). The equation of motion during the single support phase is expressed as

$$\ddot{x}_g = \frac{g}{y_h} x_g \qquad (7.1)$$

where x_g is the horizontal position of the body mass from the support point of the stance leg; g is the gravitational constant; and y_h is the body height. Given the initial horizontal position $x_g(0)$ and velocity $\dot{x}_g(0)$, the trajectory of the body mass in terms of the horizontal position $x_g(t)$ and velocity $\dot{x}_g(t)$ is expressed as follows:

$$x_g(t) = x_g(0)cosh(\frac{t}{Tc}) + Tc\dot{x}_g(0)sinh(\frac{t}{Tc}) \tag{7.2}$$

$$\dot{x}_g(t) = (x_g(0)/Tc)sinh(\frac{t}{Tc}) + \dot{x}_g(0)cosh(\frac{t}{Tc}) \tag{7.3}$$

where $Tc = \sqrt{\frac{y_h}{g}}$.

A energy-like quantity called the orbital energy [5] is defined

$$E = \frac{1}{2}\dot{x}_g^2 - \frac{g}{2y_h}x_g^2 \tag{7.4}$$

A unique orbital energy corresponds to a particular motion of the body, and it is conserved during the motion.

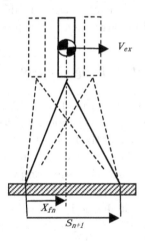

Fig. 7.5   Leg exchange condition [5].

Assume that the body is supported by only one leg at any time and the velocity of the body does not change at the support leg exchange, the value of the orbital energy E can only be changed at the support leg exchange. In fact, given the step length, s_{n+1}, at $(n + 1)$th step; the orbital energy before and after $(n + 1)$th step, E_n and E_{n+1}, respectively, the condition for support leg exchange at $(n+1)$th can be parameterized by the location of body's centre of gravity (CG) momentarily before the $(n+1)$th step, x_{fn} (with reference to the ground contact point of the support leg after nth step) (see Figure 7.5). x_{fn} can be expressed as follows [5]:

$$x_{fn} = \frac{y_n}{g s_{n+1}}(E_{n+1} - E_n) + \frac{s_{n+1}}{2}. \qquad (7.5)$$

Since orbital energy E is conserved during one walking cycle, it can be written as

$$E_n = (1/2)v_n^2 \qquad (7.6)$$

where v_n is the walking speed at the moment when $x_g = 0$ (when the body's CG is directly above the support point of the stance leg. Equation (6) can be rewritten as:

$$x_{fn} = \frac{y_h}{2 g s_{n+1}}(v_{n+1}^2 - v_n^2) + \frac{s_{n+1}}{2}. \qquad (7.7)$$

From Equation (7.7), we can see that to increase the walking speed (for a fixed step length), the bipedal robot needs to delay the support exchange moment so that x_{fn} will be larger before the support exchange.

7.4   Proposed Bipedal Walking Control Architecture

In the Linear Inverted Pendulum model, it is assumed that the legs have no inertia. However, for a physical robot, the leg's inertia is usually not negligible. Hence, the motion of the body's CG may not be well represented by Equation (7.2). Also, when the body is controlled to move along a path parallel to the ground, the orbital energy computed by Equation (7.4) may not be constant during each support phase. This is further complicated by the fact that the velocity of the body will most likely be changed during the support leg exchange due to impact forces. Such deviations from the ideal model will affect walking stability if the ideal model is used.

The walking stability of a biped is mainly determined by the swing leg behavior [2]. In this research, a Fuzzy Logic Control (FLC) based algorithm is proposed to generate the key swing leg parameters for the bipedal walking control. The Linear Inverted Pendulum (LIP) model is used to generate the initial structure for the Fuzzy Logic Controller (FLC). To tune the FLC, genetic algorithm is employed. In the following subsection, a description of the overall bipedal walking control algorithm will be given.

The next subsection highlights the intuitions concerning bipedal walking control which are obtained by looking at LIP model. The proposed FLC structure and the Genetic Algorithm (GA) will be presented in the following subsections.

7.4.1 Bipedal Walking Algorithm

This subsection provides a brief description of the overall bipedal walking control algorithm. Let's focus on the sagittal plane motion of the bipedal walking. The stance leg algorithm is mainly used to achieve the motion according to the linear inverted pendulum model, i.e.:

- to maintain a constant body height;
- to maintain an upright posture.

The swing leg algorithm is used to execute a swing foot motion which results in stable walking behavior. The key parameters for the swing foot trajectory are the lift height, swing time and the end position of the swing foot. The swing foot trajectory can be further divided into three segments, viz.: lift, swing forwards, land. Third order polynomial functions are used to describe the motion of the foot along each segment. The lift height is set to be constant to simplify the problem. So, for a given step length, the swing time becomes the only key parameter which needs to be determined. This parameter will be generated by a Fuzzy Logic Controller as described in the next subsection.

A is used to control the state transitions of the system. In this algorithm, the bipedal walking cycle is divided into four states:

- right support phase (RS);
- right-support-to-left-support transition phase (RS-TO-LS);
- left support phase (LS); and
- left-support-to-right-support transition phase (LS-TO-RS).

7.4.2 Intuitions of Bipedal Walking Control from Linear Inverted Pendulum Model

A few intuitions of bipedal walking can be obtained from the Linear Inverted Pendulum (LIP) model. These intuitions will enable the formulation of the Fuzzy Logic Controller (FLC) structure with less randomness and hence, increases the success rate for such an approach. For example, based on Equation (7.4), the initial conditions in terms of the position and velocity of body's CG are chosen to be the inputs to the FLC as they are relevant to the orbital energy of the biped. As shown in [5], the orbital energy is a parameter which allows one to check whether the biped has sufficient energy to move forwards.

By using Equations (7.2) and (7.7) and setting $x_g = x_{fn}$, the swing time can be expressed as a function of the step size at $(n+1)$th step and $v_{n+1}^2 - v_n^2$ (as shown by the surface plot in Figure 7.6). From the plot, it is observed that for a fixed value for $v_{n+1}^2 - v_n^2$, the swing time increases with the desired step size. And for a fixed step size, the swing time increases when the biped wishes to obtain a larger speed increase.

7.4.3   *Fuzzy Logic Controller (FLC) Structure*

In this subsection, the Fuzzy Logic Controller (FLC) structure is developed. The intuitions derived earlier from the Linear Inverted Pendulum (LIP) model will be used. According to the LIP model, the key variables involved in bipedal walking control are the initial position and velocity of the body's CG after the previous support exchange; the step length; the desired change in walking speed after the support exchange and the swing time.

To reduce the complexity of the problem, the step length and the walking speed are set to be constant throughout the walking task. Hence, the initial position and velocity of the body's CG are chosen to be the inputs to the fuzzy controller. In summary, the FLC takes in the initial position and velocity; and outputs the swing time for the swing leg control.

The geometric constraints of the robot and the characteristics of the linear inverted pendulum model will determine the appropriate universe of discourse for the key parameters. In this project, a simulated seven-link planar biped is used as the test-bed for the algorithms. It is a simulation model of a physical planar biped called Spring Flamingo (see Figure 7.1) which was constructed in MIT Leg Laboratory. This robot is constrained to move in the sagittal plane. Each leg has hip, knee and ankle joints which are rotary. The joint axes are perpendicular to the sagittal plane. The specifications of this biped are tabulated in Table 7.1.

Based on the specifications of Spring Flamingo, the universe of discourse for the initial body's CG position, $X_{CG}(0)$, is chosen to be within the range of [-0.24, 0] m, and that of the velocity, $V_{CG}(0)$, to be [0, 1.6] m/s. The range of the swing time, T_{Swing}, is chosen to be within [0, 0.6] s.

The universe of discourse of each variable is further divided into subsets. For input variables, 3 MFs are assigned, namely, "S" for small, "M" for

Table   7.1   Specifications   of   Spring Flamingo.

Physical parameter	Value
Total mass	$14.2kg$
Body mass	$12.0kg$
Hip to body center of mass	$0.20m$
Body moment of inertia	$0.10kgm^2$
Upper leg mass	$0.46kg$
Upper leg moment of inertia	$0.13kgm^2$
Upper leg length	$0.42m$
Lower leg mass	$0.31kg$
Lower leg moment of inertia	$0.0095kgm^2$
Lower leg length	$0.42m$
Foot mass	$0.35kg$
Foot moment of inertia	$0.0014kgm^2$
Foot height	$0.04m$
Foot length forward	$0.17m$
Foot length back	$0.06m$

medium, "L" for large. The output variable has two additional MFs, "VS" for very small and "VL" for very large. The MFs are chosen to be triangular and trapezoidal. As each input variable has 3 MFs, there are up to 9 rules. Those rules can be represented by a decision table as shown in Table 7.3. Figure 7.6 summarizes the FLC structure.

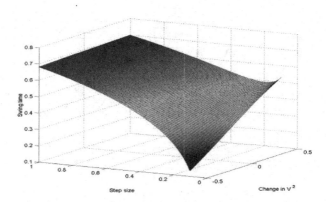

Fig. 7.6   Surface plot of swing time (for $E > 0$) with respect to change in velocity (squared) and step length.

7.4.4 *Genetic Algorithm Implementation*

When applying Genetic algorithm (GA) to FLC, there are two basic decisions to be made, viz.: how to code the possible solutions to the problem as a finite parameter string, and how to evaluate the fitness of each string. This subsection will first present the coding scheme followed by the evaluation criteria. After that, the detailed description of the GA operators will be presented.

7.4.4.1 *Coding the information*

The information of FLC to be coded is divided into that belongs to the membership functions (MFs) and that belongs to the fuzzy rules. The former contains a set of parameters representing the bases of the membership functions, whereas the latter encodes the fuzzy control rules.

Once the structure of the FLC is determined, the search space for each variable can be partitioned as shown in Figure 7.7(a). It can be seen that to represent all the fuzzy sets in a variable with 3 MFs, 3 parameters limited by the search margins (lower bounds and upper bounds) will be needed. Similarly for 5 MFs, 5 parameters are needed. Thus, in total, 11 parameters are used to represent all MFs in the proposed FLC. This set of parameters will be coded into an array of 11 real numbers which constitutes the "MF Gene" (Figure 7.7(b)).

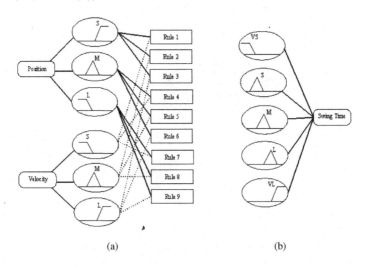

(a)   (b)

Fig. 7.7 Proposed FLC structure.

In this FLC system, the output has 5 MFs. Each of which is assigned an integer number as shown in Table 7.2. The fuzzy rules represented by the decision table (Table 7.3) can then be directly translated into a string of 9 integer numbers ($\in [0,4]$) called the "Rule Gene" (see Figure 7.8).

Table 7.2 Mapping of output fuzzy sets to integers.

MF	Code
VS	0
S	1
M	2
L	3
VL	4

Table 7.3 FL decision table.

		Pos.		
		S	M	L
Vel.	S	M	L	VL
	M	S	M	L
	L	VS	S	M

7.4.4.2 *Evaluation*

To evaluate each individual, an appropriate fitness function is needed. Since the main interest is on the success of bipedal walking, the number of steps that the biped can walk without falling can be used to evaluate the performance of each individual. Thus, one possible fitness function is:

$$\text{Fitness 1} = \text{Number of successful walking steps} = S_{tot} \qquad (7.8)$$

This fitness function is simple. However, it does not take care of the other key walking parameters like velocity, swing time, etc. Also, the walking period may be erratic unless penalty is given to such behavior. To achieve a more periodic walking motion, another component which penalizes the swing time fluctuation can be added into the fitness function. This results in the following fitness function:

$$\text{Fitness 2} = S_{tot} - P_1 \qquad (7.9)$$

where P_1 = penalty associated with erratic swing time = $\alpha \sum_{n=2}^{N} (t_s^{(n)} - t_s^{(n-1)})^2$, $t_s^{(n)}$ is the actual swing time for the nth step, N is the total steps

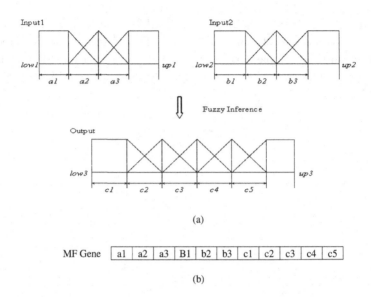

Fig. 7.8   Encoding the MF information: (a) Partition of universe of discourse; (b) A sample representation of MF Gene.

successfully walked and α is the weighting coefficient.

7.4.4.3   *Evolutionary operators*

Figure 7.9 shows the structure of the individuals in a population of size n. The major components of a chromosome are the fitness value, the MF and the Rule genes. This subsubsection explains how the genetic operators are applied to this chromosome structure.

- Reproduction: According to the fitness value of each individual, first apply the Roulette Wheel selection method, where members with larger fitness values receive a higher possibility to be reproduced in the new generation.
- Crossover: Since each chromosome has rule gene (consists of integer numbers) and MF gene (consists of real numbers), rule genes will crossover with rule genes and MF genes with MF genes. Figure 7.10 shows an illustration of the crossover process. To maintain the diversity of the population and prevent premature convergence, the crossover rate should be high.
- Mutation: After a crossover is performed, mutation takes place. The

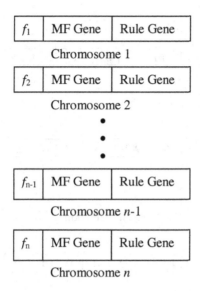

Fig. 7.9   Structure of individuals in a population information.

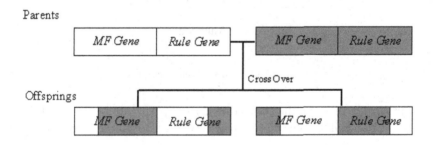

Fig. 7.10   MF and Rule Gene crossover.

mutation is intended to prevent solutions from being trapped in local optima. The mutation operation is used to randomly changes the offspring resulted from crossover. Again, the mutation is divided into two different parts: rule gene mutation and membership function gene mutation.

Rule gene mutation is performed by randomly selecting an element in the rule gene and replacing it by any integer number in $[0, 4]$. MF mutation is performed by randomly selecting an element in the MF gene and multiplying it by a real number in $[0.8, 1.2]$. Figure 7.11 illustrates the mutation process.

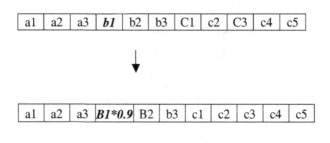

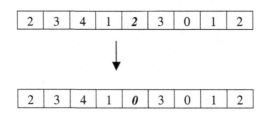

(a) MF gene Mutation

(b) Rule gene Mutation

Fig. 7.11   MF and rule genes mutation.

- Elitism: To preserve the best individuals, top 5% chromosomes in the old generation will be selected to replace the worst 5% in the new generation.

7.5   Simulation Result

This section describes the simulation result of the walking control implementation of the simulated bipedal robot (Spring Flamingo) based on the proposed GA tuned FLC walking algorithm. Yobotics! Simulation Construction Set is used to simulate the bipedal robot while its joints are given respective torques generated by the walking control algorithm. It takes care of the dynamic behavior of the robot and also allows ground contact formulation. Beside the robot simulator, a fuzzy engine and genetic algorithm software packages are also used for fuzzy evaluation and GA process, respectively.

In the simulation, the step length is set at 0.4 m. The initial body's CG position and velocity are set at -0.1667 m and 0.975 m/s, respectively. The fitness function 2 (Equation (7.9)) is used in the GA process. The objective is to maximize the number of steps that the robot can walk as well as trying to keep the walking cycle to be more regular (less erratic). α of the fitness function is chosen to be 50.

The GA process starts with an initial population of size 50, each with random initialization of MF and Rule genes. The first as well as the most important objective is to find an elite chromosome with suitable settings for MF and Rule genes such that the FLC will enable continuous stable and regular walking.

After around 25 generations, the best chromosome achieves continuous walking throughout the allowed time (100 s). The highest fitness values of each generation, i.e. the maximum number of steps (with step length set to 0.4 m) walked within 100 s, are plotted in Figure 7.12.

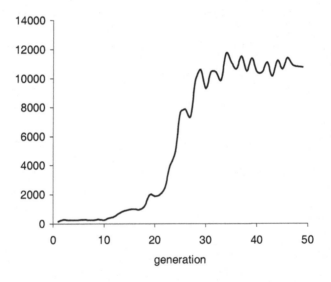

Fig. 7.12   Total fitness over generation (with Fitness 2) information.

Figure 7.13 shows an example of a successful FLC solution. For this FLC, the sequence of swing time values generated during the walking implementation is captured in Figure 7.14. The image sequences of the successful walking implementation are as shown in Figure 7.15.

7.6   Summary

This chapter proposes a GA tuned FLC algorithm which generates the swing time for walking control of a biped. The simulation results showed that the GA is effective in tuning the FLC parameters. After evolutionary optimization using GA, the FLC is able to generate a sequence of swing times which results in continuous walking motion.

This work has shown the feasibility of the proposed GA-FLC approach for bipedal walking control. The approach can be extended further by increasing the input dimension of the FLC so that parameters such as the step length and the walking velocity can be included. The GA evaluation criteria can also be modified to include other specifications like walking speed. The application of GA-FLC algorithm to bipedal walking control on uneven terrains will also be studied in the future.

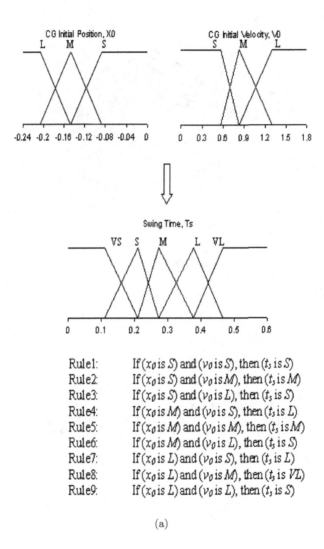

Rule1:   If $(x_0$ is $S)$ and $(v_0$ is $S)$, then $(t_s$ is $S)$
Rule2:   If $(x_0$ is $S)$ and $(v_0$ is $M)$, then $(t_s$ is $M)$
Rule3:   If $(x_0$ is $S)$ and $(v_0$ is $L)$, then $(t_s$ is $S)$
Rule4:   If $(x_0$ is $M)$ and $(v_0$ is $S)$, then $(t_s$ is $L)$
Rule5:   If $(x_0$ is $M)$ and $(v_0$ is $M)$, then $(t_s$ is $M)$
Rule6:   If $(x_0$ is $M)$ and $(v_0$ is $L)$, then $(t_s$ is $S)$
Rule7:   If $(x_0$ is $L)$ and $(v_0$ is $S)$, then $(t_s$ is $L)$
Rule8:   If $(x_0$ is $L)$ and $(v_0$ is $M)$, then $(t_s$ is $VL)$
Rule9:   If $(x_0$ is $L)$ and $(v_0$ is $L)$, then $(t_s$ is $S)$

(a)

Fig. 7.13   (a) MF and Rule of FLC with GA (b) the surface plot of the swing time.

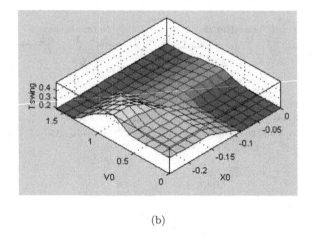

(b)

Fig. 7.13   (*Continued*)

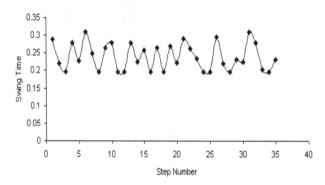

Fig. 7.14   Swing time for a sequence of steps (with Fitness 2).

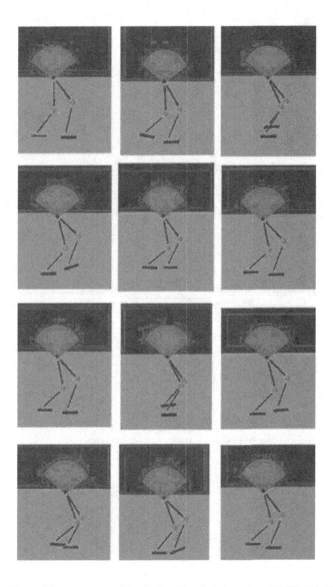

Fig. 7.15 The walking sequence of the Spring Flamingo based on GA-FLC (with Fitness 2) implementation. The images are 0.15s apart.

Bibliography

[1] M. R. Akbarzadeh-T. and E. Tunstel and K. Kumbla and M. Jamshidi, Soft computing paradigms for hybrid fuzzy controllers: experiments and applications, 1998.

[2] C. M. Chew, Dynamic Bipedal Walking Assisted by Learning, *Robotica*, vol. 20, pp. 477–491, 2002.

[3] O. Cordon and F. Herrera, A two-stage evolutionary process for designing TSK fuzzy rule-based systems, *IEEE Trans. on Syst., Man, Cybern.*, pt, B, vol. 29, pp. 703–715, December 1999.

[4] D. E. Goldberg, *Genetic Algorithms in Search, Optimization and Machine Learning*, Kluwer Academic Publishers, Boston, MA, 1989.

[5] S. Kajita, T. Yamaura, and A. Kobayashi, Dynamic walking control of a biped robot along a potential energy conserving orbit, *IEEE Transactions on robotics and automation*, vol.8, no. 4, August, 1992.

[6] S. J. Kang, H. S. Hwang, and K. N. Woo, Evolutionary design of fuzzy rule base for nonlinear system modeling and control, *IEEE Trans. on Fuzzy Syst.*, vol. 8, pp. 37–45, February 2000.

[7] W.T. Miller, Learning dynamic balance for a biped walking robot, *Proceedings of the IEEE Int. Conf. on Neural Networks*, June, 1994.

[8] F. Miyazaki and S. Arimoto, A control theoretic study on dynamical biped locomotion, *ASME J. Dynam.Syst.Meas.Contr.*, vol.102, pp.233–239, Dec,1980.

[9] M. H. Raibert, Running with symmetry, *Inc. J. Robotics Res.*, vol.5, no.4, pp. 3–19, 1986.

[10] M. Russo, Genetic fuzzy learning, *IEEE Trans. on Evolutionary Computation*, vol. 4, no.3, pp. 259–273, 2000.

[11] Y. Shi, R. Eberhart, and Y. Chen, Implementation of evolutionary fuzzy systems, *IEEE Trans. on Fuzzy Syst.*, vol. 7, no. 2, pp.109–119, 1999.

[12] C. L. Shih, and C. J. Chiou, The motion control of a statically stable biped robot on an uneven floor, *IEEE Trans. Syst., Man, Cybern.*, vol. 28, no.2, 1999.

[13] P. Thrift, Fuzzy Logic Synthesis with Genetic Algorithms, *Proceedings 4th. International conference on Genetic Algorithms*, pp. 509–513. Morgan

Kaufmann, 1991.

[14] M. Vukobratovic, B. Borovac, D. Surla, D. Stokic, Biped locomotion: Dynamics, stability, control and application, *Scientific fundamentals of robotics 7*, Springer–Verlag, New York, 1990.

[15] L. Zadeh, Outline of a New Approach to the Analysis of Complex Systems, *IEEE Trans. on Sys., Man and Cyb.* 3, 1973.

Chapter 8

Bipedal Walking: Stance Ankle Behavior Optimization Using Genetic Algorithm

This chapter studies the application of Genetic Algorithm (GA) as an optimization tool to search and optimize key parameter in the walking controlling of a humanoid robot. In this study, Virtual Model Control is employed as a control framework where ankle gain plays an important part in regulating forward velocity during walking. Optimal value of the ankle gain is searched and optimized by GA such that stable walking gait with smooth velocity profile can be achieved. Simulation results show that the walking gait is stable and its forward velocity profile has lower variation than those before optimization. The results are verified by comparing with the results obtained by enumerative method of optimization. The effects of GA's parameters on simulation results will also be discussed.

8.1 Introduction

Nowadays, realizing a stable walking gait for a humanoid robot is no longer a very tough task, thanks to a great deal of control algorithms developed so far. However, these algorithms are tedious to implement due to the requirement for manual parameter-tuning. The process can be time consuming and the walking gait is usually not optimized. Some well-known bipedal walking control algorithms include ZMP-based method [1, 2], Virtual Model Control [3], Linear Inverted Pendulum [4], and inverse-kinematics based method [5], etc. The first algorithm is used extensively by many researchers in Japan. This method controls the biped based on the planning of ZMP (Zero Moment Point) trajectories. Virtual model control is a control framework which controls the biped by transforming virtual forces into joint torques. The virtual forces are generated by virtual components placed at strategic positions. Linear inverted pendulum is a control method that controls the biped by constraining the center of mass to move along a constraint line. In inverse kinematics planning method, the trajectories are planned in Cartesian space and will be transformed to joint space trajectories for realization using inverse kinematics algorithm. All these algorithms introduce many coefficients that need to be tuned intensively until the biped can walk. It is apparent that the tuning processes of these approaches are not effective. There is no unified method to design those parameters as in PID control. Most are tuned manually. In the lack of such effective design method, a searching algorithm is usually needed. In this chapter, we present an approach that uses Virtual Model Control as the control framework and GA as a searching algorithm to search for the optimal value of stance ankle joint's control gain. Although Virtual Model Control has also other important parameters which require tuning, the tuning process for these parameters can be intuitively done. We focus on the stance ankle as it plays an important role in the stability of the bipedal walking and is directly related to forward velocity as shown in later section. This chapter is organized as follows. Section two will briefly review Virtual Model Control and explain the effect of stance ankle torque on walking speed. In Section three, the application of GA to search for the optimal value of ankle gain will be presented. Section four presents the simulation results and discussion. Finally, the conclusion will be made in Section five.

8.2   Virtual Model Control

Virtual model control [3] is a control framework for multi-body system. In this framework, virtual components such as spring damper, etc., are placed at strategic positions to achieve certain control objective. The virtual components will generate corresponding virtual forces based on their constitutive properties. These forces are then transformed into joint torques which will create an effect as if the virtual forces exist. One of the advantages of this algorithm is its ease of implementation. We can plan the motion of the system in the Cartesian space, which are more intuitive, instead of the joint space. For more details on Virtual Model Control, refer to [3]. In this section, some of its key points will be presented. For the case of a biped, a virtual rotational spring and damper is attached to the body to generate a virtual moment f_θ. This virtual moment will maintain the body's pitch angle at upright position. Each leg of the biped has two phases: stance and swing. In the stance phase, a virtual spring and damper are mounted vertically on the hip to generate the vertical virtual force f_z that can maintain the hip or body height. In the swing phase, the ankle of the leg is mounted with a vertically placed spring-damper pair and a horizontally placed damper to generate the vertical and horizontal virtual forces, f'_z and f'_x, respectively, to maintain the swing ankle's height and forward speed, respectively. Fig. 8.1 shows the mounting of the virtual components on the body, the stance and the swing leg.

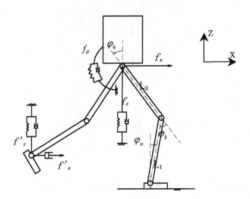

Fig. 8.1   The placement of virtual components on the biped.

The specifications of the simulated biped are taken from a physical biped called NUSBIP which was developed in our laboratory. Fig. 8.2 shows the photograph of NUSBIP. Its specifications are summarized in Table 8.1. The desired walking parameters are presented in Table 8.2.

Table 8.1   Specifications of the simulated biped.

Parameters	Value
1. Body mass	10.45kg
2. Thigh mass	2.68kg
3. Shank mass	1.68kg
4. Foot mass	1.52kg
5. Thigh length L_2	32cm
6. Shank length L_1	32cm
7. Foot length	23cm
8. Foot thickness	5.5cm

Assume that the virtual forces $[f_x, f_z, f_\theta]^T$ as shown in Fig. 8.1 have been generated by appropriate virtual components. They can be virtually

Fig. 8.2   The photograph of NUSBIP.

Table 8.2   Walking parameters.

Parameters	Value
1. Body height	65cm
2. Step length	35cm
3. Desired speed	0.75m/s

realized by applying the corresponding ankle, knee and hip joints' torques of the stance leg, $[\tau_a, \tau_k, \tau_h]^T$:

$$
\begin{bmatrix} \tau_a \\ \tau_k \\ \tau_h \end{bmatrix} = \begin{bmatrix} -L_1c_a - L_2c_{a+k} & L_1s_a - L_2s_{a+k} & -1 \\ -L_2c_{a+k} & -L_2s_{a+k} & -1 \\ 0 & 0 & -1 \end{bmatrix} \begin{bmatrix} f_x \\ f_z \\ f_\theta \end{bmatrix}
$$

$$
\equiv \begin{bmatrix} A & B & -1 \\ C & D & -1 \\ 0 & 0 & -1 \end{bmatrix} \begin{bmatrix} f_x \\ f_z \\ f_\theta \end{bmatrix} \tag{8.1}
$$

where f_x, f_z and f_θ are the horizontal, vertical and rotational virtual force applied on the body of the biped (at the hip); $c_a = cos(\varphi_a)$; $c_a + c_k = cos(\varphi_a + \varphi_k)$; $s_a = sin(\varphi_a)$; $s_a + s_k = sin(\varphi_a + \varphi_k)$; $A = -L_1c_a - L_2c_{a+k}$;

$B = -L_1 s_a - L_2 s_{a+k}; \ C = -L_2 c_{a+k}; \ \text{and} \ D = -L_2 s_{a+k}.$

In the implementation, f_z and f_θ are generated by the virtual components attached to the body to maintain the body's height and pitch angle, respectively. To maintain the horizontal walking speed, instead of prescribing a horizontal component to the biped, we propose a direct stance ankle joint control law to achieve the horizontal walking speed control. For humanoid or bipedal robot, the foot is not fixed onto the ground. As a result, if the resultant stance ankle torque derived from the prescribed control law is too high, the foot may rotate and this will lead to unpredictable result. Therefore, the ankle control law must be properly chosen. In this work, the stance ankle control law is formulated as follows:

$$\tau_a = K(v_{desired} - v) \tag{8.2}$$

where v is current forward velocity of the body; $v_{desired}$ is the desired forward velocity; and K is the control gain whose optimal value will be searched using GA. From Eq. (8.1), the stance ankle torque can be expressed as follows:

$$\tau_a = A f_x + B f_z - f_\theta. \tag{8.3}$$

Thus,

$$f_x = \frac{\tau_a + f_\theta - B f_z}{A}. \tag{8.4}$$

Substituting Eq. (8.4) into the hip and knee torque equations, we have:

$$\tau_k = \frac{C}{A}\tau_a + (D - \frac{BC}{A})f_z + (\frac{C}{A} - 1)f_\theta \tag{8.5}$$

$$\tau_h = -f_\theta. \tag{8.6}$$

In summary, in the stance phase, the joints' torques τ_a, τ_k and τ_h are controlled based on Eq. (8.2), (8.5) and (8.6), respectively. For the stance knee torque in Eq. (8.5) to be determined, the denominator $A = -L_1 c_a - L_2 c_{a+k}$ must be non-zero. For the case of $L_1 = L_2$, A equals to zero when $\varphi_k + 2\varphi_a = 180^\circ$ or $\varphi_k = \pm 180^\circ$. The first condition only happens when the hip is at the same level as the ankle joint as demonstrated in Fig. 8.3. The second condition happens when the knee is completely flexed such that the thigh and shank coincide. However, these conditions may not happen during walking. Hence, A is always non-zero during walking.

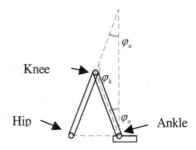

Fig. 8.3   One of the stance leg configuration when A=0.

For the swing leg, the control of the joints is easier because all the virtual forces can be set independently. The Jacobian for the swing leg is slightly different from that of the stance leg because the reference frame for the derivation is different:

$$
\begin{bmatrix} \tau'_h \\ \tau'_k \\ \tau'_a \end{bmatrix} = \begin{bmatrix} -L_1 c_{h+k} - L_2 c_h & L_1 s_{h+k} + L_2 s_h & 1 \\ -L_1 c_{h+k} & L_1 s_{h+k} & 1 \\ 0 & 0 & 1 \end{bmatrix} \begin{bmatrix} f'_x \\ f'_z \\ f'_\theta \end{bmatrix} \tag{8.7}
$$

8.3   Genetic Algorithm (GA)

8.3.1   *GA's Operations*

Genetic algorithm is inspired by the selection process that governs the evolution of all creatures. During this process, the fitter individuals have higher chance to survive and the less fit ones are gradually phased out. The superior individuals are then mated to produce better and better generations. All creatures in our planet evolve in this manner and have been adapting to the external environment very well. This fact has inspired many researchers to adopt such mechanism to solve optimization problem by treating each solutions as individuals. The solution candidates evolve from generations to generations until stopping criteria are met. GA has been applied successfully in a wide range of fields, from biology, medicine, computer science, engineering to social science [6]. In GA, all the parameters are coded into a string of symbols, usually binary digits {0, 1}. The coded string is called an individual or chromosome. Initially, a population of individuals is generated randomly. GA will test each individual in the population and return an associated fitness value. Based on these fitness values, GA's operations will be performed to generate a new generation which will contain fitter individuals. These operations include reproduction, crossover and mutation. During the reproduction phase, the individuals with higher fitness values will have higher chance to be reproduced. To perform this operation, each individual is fixed to a slot in a roulette wheel. The size of the slots is proportional to the fitness value of the individuals. After each spinning of the wheel, it is obvious that the individuals fixed to larger slots will have higher chance to be selected. After the reproduction, the individuals in the new generation will be mated randomly to perform crossover operation in which two individuals will exchange their "genes", which is a portion of the coded string. The crossover operation is illustrated in Fig. 8.4.

Fig. 8.4   Crossover operation (Adapted from [6]).

The mutation operation occasionally modifies some specific position in the string to generate new solution. This operation may generate better individual, but may also generate worse individual. Therefore, the probability for this operation to happen has to be low. One advantage of using GA over other algorithms is that this technique performs the solution search at different locations simultaneously in the searching space rather than only one point. This helps to overcome the problem of getting stuck at local optima. The other advantage of GA is that this algorithm performs the search using the fitness value as the only information. It does not require any function differentiation, so the problem of discontinuity does not happen here.

8.3.2   GA's Parameters

In this study, GA is used to search for the optimal value of the ankle gain. This gain is coded into a ten-bit binary number. We wish to search for an optimal value within the range from 0 to 15. This range is selected because the gain above 15 always causes the bipedal walking to fall. Therefore, the resolution of the ankle gain is $15/2^{10}$. The percentage of crossover is selected to be 90%. The rate of mutation is zero. The parameters of GA are summarized in Table 8.3.

GA's parameters	Value
1. Crossover rate	90%
2. Mutation rate	0%
3. Number of individual in a population	20
4. Number of generation	30

8.3.3   *Fitness Function*

Fitness function is one of the most important factors of this searching algorithm. This is the only basis for the algorithm to perform the search operation. The fitness value must reflect the objective that we want to achieve. In this simulation, we want to find the value of ankle gain such that the horizontal velocity profile of the body is smooth and the average velocity is close to the desired velocity. So, the performance index must consist of both the average amplitude of the velocity profile and the difference between average speed and desired speed. The performance index is as followed:

$$P = \frac{1}{n}\sum_{i=1}^{n}(V_{max} - V_{min}) + |V_{avg} - V_{desired}| \qquad (8.8)$$

where n is the number of walking steps; i denotes the ith step; V_{max} and V_{min} are the maximum and minimum walking speed, respectively, within the ith step; V_{avg} is the average walking speed of the biped; and $V_{desired}$ is the desired walking speed of the biped. The objective is to find a gain for the ankle torque equation such that this performance index is minimized. In our simulation, we have converted this performance index into a fitness function by taking its inverse. GA will try to maximize this fitness function. The fitness function is written as follows:

$$F = \frac{1}{P} = \frac{n}{\sum_{i=1}^{n}(V_{max} - V_{min}) + n|V_{avg} - V_{desired}|}. \qquad (8.9)$$

The above fitness function is only evaluated for the gain value which is able to make the biped walks stably. However, there are some gain values which result in the biped falling down while walking. In this case, the fitness function is set to zero so that those gain values are left out in the next generation.

8.4 Simulation Results and Discussion

8.4.1 *Convergence to Optimal Solution*

The simulation is done in Yobotics *, a dynamics simulation software, which allows the simulation trials to be run in batches. After each simulation trial, the fitness values and the corresponding ankle gains of the individuals are recorded in a data file. The fitness values of 30 generations are shown in Fig. 8.5 and the ankle gain of all 600 individuals is shown in Fig. 8.6. The convergence occurs after 14 generations.

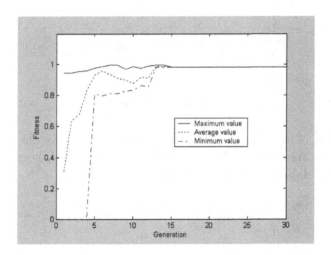

Fig. 8.5   Fitness values of 30 generations.

The range of ankle gain's values shrinks after a few generations and finally converges to a single value. In the first generation, the ankle gains are distributed randomly spanning a wide range of value. This generation usually contains some values that make the biped fall down quickly. These values of ankle gain are called defective individuals. The fitness values of defective individuals are set to zero such that they have no chance to be reproduced to the next generations. However, defective individuals keep appearing in the next few generations as shown on Fig. 8.5 despite of the above filtering mechanism. This phenomenon suggests that the mating of two superior individuals does not guarantee a production of the two new

---

*Developed by Yobotics, Inc., http://yobotics.com

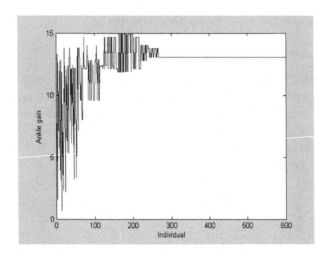

Fig. 8.6   The ankle gain of all individuals.

individuals of higher quality. However, the probability for this to happen is very low. As shown in Fig. 8.7, the defective individuals keep decreasing and finally disappear in further generations.

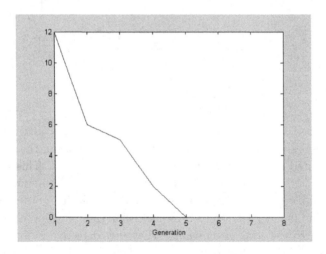

Fig. 8.7   The decline of the number of defective individuals over generations.

Velocity profiles of the biped with respect to some values of the ankle

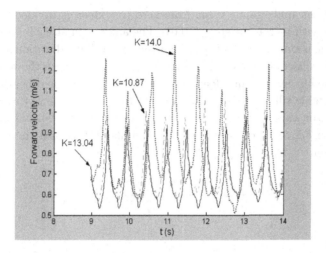

Fig. 8.8   The velocity profiles for the case of optimal ankle gain K=13.04 compared with those of K=10.87 and K=14.

gain are shown in Fig. 8.8. The optimal ankle gain is $K = 13.04$ which produces a more uniform and lower variation velocity profile than those of non-optimal values. Even though some non-optimal values of ankle gain are able to make the biped to walk longer than the time limit set for the simulation, their derived bipedal motion are jerkier, more oscillated and less natural than the one produced by the optimal value. Therefore, it is much better to use a systematic optimization tool rather than a manually tuning of the parameters which is usually stopped at the values resulted in stable but non-optimal gaits.

8.4.2   A Comparison with Solution Produced by Enumerative Method of Optimization

Among the optimization tools,enumerative method is the most costly one. It has to examine every point in the searching space and selects the optimal solution. However, the results obtained from this method are reliable because all possible values of the parameters are evaluated and compared. Therefore, this method is used to verify the results obtained by GA. The enumerative method should only be used in low dimensional space because it is extremely ineffective. In our study, the time required to run the simulation using this method is almost three-fold that of GA. In this experiment, the range of ankle gain's value from zero to sixteen is divided into small divisions with increment step of 0.01. The ankle gain at each point within that range is simulated, and the corresponding walking duration and fitness value are collected. Some values of ankle gain cannot keep the biped walking stable but make it fall down after a few seconds; some others can make it walk stable till the end of the simulation. Fig. 8.9 shows the durations a biped can walk with corresponding gain's value. It is shown that ankle gain from 8.5 to 15 can make the biped walk stably until the end of the simulation.

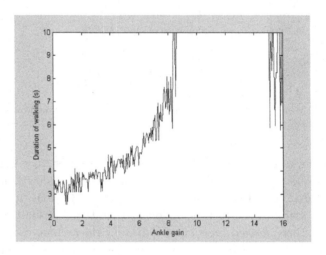

Fig. 8.9   Duration that the biped can walk with respect to ankle's gain value.

The fitness values corresponding to gain values within the range from 0 to 16 are shown in Fig. 8.10. From this graph we can see that the optimal value is 13.1 with corresponding fitness value of 1.0113. The optimal solution obtained by GA is 13.04 and 0.9852 for ankle's gain value and fitness value, respectively. These results show that GA is quite accurate and much more effective than enumerative method.

8.4.3   *The Effects of GA's Parameters*

In the above discussion, the simulation is done with 90% crossover rate. In an attempt to examine the effect of such parameter, we have tried several other values. The results show that when the crossover rate is increased,

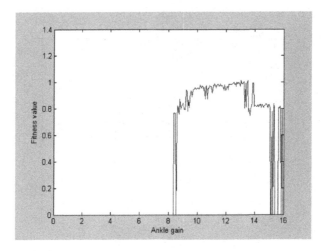

Fig. 8.10   Fitness values vs. ankle gain's value.

GA will take longer to converge. Inversely, if the crossover rate is low, GA converges faster. However, in such a case the converged fitness value is usually not as good as that of higher crossover rate. Lower crossover rate means more individuals in the reproduction pool will be retained in the new generation. This will make GA converge faster, but the problem of local optima will arise. Fig. 8.11 shows the fitness value for the simulation batch with 80% crossover rate. The convergence rate is also affected by the initial population. The initial population in GA is generated randomly, therefore it is quite possible that all the individuals of the population are bad or all are good or a mixture of both. The bad individual is the one that cannot make the biped walk stably. If the initial population contains all bad individuals, GA may not converge. If there are more good individuals in initial population, the convergence rate is faster. To increase the number of good individuals in the initial population, the size of the population must be large enough.

In summary, the simulation results have shown that GA is an effective search and optimization tool for such a simple problem. Furthermore, the implementation demonstrates an automated tuning process which requires less effort than the manual tuning process. GA differs from a random search algorithm despite it bases on some randomness in making decision. They exploit past experiences to evolve by combining the building blocks of the good individuals. This simple problem allows us to gain some insights on

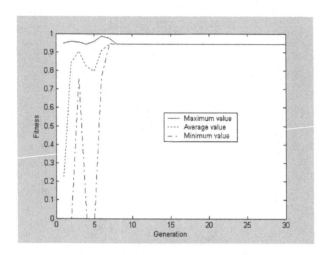

Fig. 8.11   The convergence of fitness values when the crossover rate is 80%.

GA implementation issues. With these insights, it is easier to extend the implementation to more complex problems pertaining to bipedal walking control. In future studies, the problem will be extended to the search of more than one parameter. Different fitness function will also be explored, especially for more complex problem, as the search space will be larger than that of one parameter.

8.5   Summary

In this paper, we have presented the application of GA to search for the optimal value of the stance ankle joint control gain. The simulation results show that the fitness value tends to increase over generations, and after 14 generations it converges to a single value. The converged ankle gain value results in a smooth motion and the average speed being close to the desired speed. This result has been verified by comparing with the result produced by the enumerative method of optimization. In the future studies, multi-variable optimization problem for bipedal walking will be explored. New fitness functions that may lead to different walking gaits will also be studied.

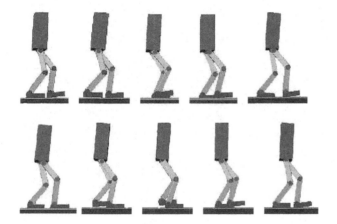

Fig. 8.12   One walking cycle of the simulated biped.

Bibliography

[1] K. Mitobe, G. Capi, Y.Nasu, Control of Walking Robot Based On Manipulation Of The Zero Moment Point, *Robotica (2000)*, Vol. 18, pp. 651–657.

[2] T. Sugihara, Y. Nakamura, H. Inoue, Realtime Humanoid Motion Generation Through ZMP Manipulation Based On Inverted Pendulum Control, *Proceedings Of The 2002 IEEE International Conference On Robotics And Automation*, Washington DC, May 2002.

[3] J. Pratt, C.-M. Chew, A. Torres, P. DilWorth, G. Pratt, Virtual Model Control: An Intuitive Approach For Bipedal Locomotion, *The International Journal Of Robotics Research*, Vol. 20 No. 2 February 2001, pp. 129–143

[4] S. Kajita, K. Tani, Experiment Study of Biped Dynamic Walking, *IEEE International Conference On Robotics And Automation*, Nagoya, Japan, May, pp. 21–27.

[5] Q. Huang, K. Yokoi, S. Kajita, K. Kaneko, H. Arai, N. Koyachi, and K. Tanie, Planning Walking Patterns for a Biped Robot, *IEEE Transactions on Robotics and Automation*, Vol. 17, No. 3, June 2001.

[6] D.E. Goldberg, *Genetic Algorithm in Search Optimization and Machine Learning*, Addison-Wesley, MA, 1989.

[7] G. Capi, S. Kaneko, K. Mitobe, L. Barolli, Y. Nasu, Optimal Trajectory Generation For A Prismatic Joint Biped Robot Using Genetic Algorithms, *Robotics And Autonomous Systems 38 (2002)*, pp. 119–128.

[8] D. Golubovic, H. Hu, A Hybrid Evolutionary Algorithm for Gait Generation of Sony Legged Robots, *28th Annual Conference of the IEEE Industrial Electronics Society*, Sevilla, Spain, November 5–8, 2002.

[9] F. Yamasaki, K. Endo, H. Kitano, M. Asada, *Acquisition of Humanoid Walking Motion Using Genetic Algorithm- Considering Characteristics Of Servo Modules*, Technical Report.

[10] T. Shibata, T. Abe, Motion Planning By Genetic Algorithm For A Redundant Manipulator Using A Model Of Criteria Of Skilled Operators, *Information Sciences 102*, pp. 171-186 (1997).

[11] F. Yamasaki, K. Hosoda, M. Asada, *An Energy Consumption Based Control For Humanoid Walking*, Technical Report.

[12] J. Kamiura, T. Hiroyasu, M. Miki, S. Watanabe, MOGADES: Multi-Objective Genetic Algorithm with Distributed Environment Scheme, *Proceedings of the 2nd International Workshop on Intelligent Systems Design and Applications.*

most recent development of this direction are presented. The basic concept of evolvable hardware and its classification are first discussed. The evolvable hardware can be classified along the dimensions of artificial evolution and hardware devices, evolution process, adaptation methods, and application areas. In addition, the promises and challenges of its applications in evolutionary robotics are also given. In Chapter 3, a real-world implementation of autonomous robotic navigation system based on a type of evolvable hardware called FPGA (Field Programmable Gate Array) is fleshed out. The hardware-based robotic controller design including Boolean function controller, chromosome representation, evolution and adaptation methodology, and robot navigation tasks is first discussed. Then hardware configuration and development platform are introduced. Two practical implementations of the developed robotic navigation system are then detailed, which include light source following task and obstacle avoidance using a faulted mobile robot. In Chapter 4, intelligent sensor fusion and learning for autonomous robot navigation is discussed. First development platform and controller architecture are introduced, which include the Khepera robot and Webots software, hybrid control architecture, and robot functional modules. Then a Multi-Stage Fuzzy Logic (MSFL) sensor fusion system is discussed, where the issues on feature-based object recognition and MSFL inference system are addressed. The Grid Map Oriented Reinforcement Path Learning (GM-RPL) is then fleshed out. It implementation in the physical environment is also presented. In Chapter 5, Task-Oriented Developmental Learning (TODL) for humanoid robotics is discussed. First the TODL system including task representation, the AA-learning, and task partition is introduced. Then the self-organized knowledge base is given. A prototype TODL system coupled with other issues such as TODL mapping engine, knowledge database, and sample tasks is fleshed out later on. In Chapter 6, a general control architecture for bipedal walking is discussed, which is based on a divide-and-conquer approach. The sagittal-plane motion-control algorithm is formulated based on a control approach called virtual model control. A reinforcement learning algorithm is used to learn the key parameter of the swing leg control task so that stable walking can be achieved. In Chapter 7, a genetic algorithm tuned fuzzy logic controller is proposed for bipedal walking control implementation. The basic structure of fuzzy logic controller is constructed based on the linear inverted pendulum model. Genetic algorithm is implemented to search and optimize the fuzzy logic controller parameters. In Chapter 8, the genetic algorithm is used as an optimization tool to search and optimize key parameter in the walking controlling of a humanoid robot. Virtual model control is employed as a control framework where ankle gain plays an important part in regulating forward velocity during walking.

Whether or not the potentially great engineering benefits of fully autonomous intelligent robots turn out to be completely realizable in real-world applications, the fundamental groundwork developed herein must provide some meaningful information for the future development of this domain.

9.2 Future Research Emphases

Evolutionary robotics is a young and vibrant field. Although a lot of progress has been made in the recent years, many open-ended problems still exist. Below are several major concerns of future research in this field.

9.2.1 *On-Line Evolution*

The most difficult result to obtain and yet not being obtained in evolutionary robotics is in actuality the deviation of complete robotic behavior by means of on-line evolution. For a truly intelligent autonomous robot, it should be able to deal with the uncertainties occurred during its operations in an adaptive manner. Its controller should not be pre-determined, since hard-coded robotic controller is not capable of exhibiting adaptive behaviors in the presence of situations that are not considered during the controller design process. To effectively cope with the unexpected situations, the robot should be able to adjust its behavior by itself accordingly in order to adapt to the new operating conditions. Without this ability, the robot is not really intelligent and can only operate in the environment where it was trained before. However, the physical environments are usually super complex. Modeling such an actual environment is computationally expensive and robotic controller design based on such a model is very costly even if it is possible. Furthermore, it is usually not viable to model the exact operating environment precisely since it may vary during system operations. On-line evolution is a difficult problem, considering it requires the robot to be able to sense the real-world uncertainties, perform suitable decision-making, and take corresponding measures to adapt to the new system model in a very efficient fashion. It demands significant improvements in both hardware and software aspects, including elevated hardware power, more efficient algorithms, and even totally different controller design philosophies. Achieving fully autonomous robots by incorporating on-line evolution mechanism is one of the major design objectives in the future effort of this field.

9.2.2 *Inherent Fault Tolerance*

Fault tolerance is an inherent requirement for any robust autonomous robotic systems. Due to the complexity of such systems and the harsh physical environments, it is fairly possible that the system components will become out of service during operations. Thus, it is highly undesirable to produce fragile robotic systems which are not able to handle both internal and external abnormal situations. Fortunately, certain techniques discussed in this book show some promise for obtaining a certain level of fault tolerance for the autonomous robots which are prone to fail in the complex practical operating world. For instance, hardware-based robotic controller is a promising domain where robots can perform their tasks in a more reliable fashion. A commercially useful circuit must operate over a range of temperatures and power supply voltages, and be implemented on many different chips that are always slightly different to each other. The major challenge to be faced by unconstrained intrinsic evolvable hardware evolution is to exploit maximally those properties which are sufficiently stable over the required range of operating conditions to support a robust behavior, but meanwhile to be tolerant to variations in other aspects of the medium. This involves fault tolerance of intrinsic EHW. In certain kinds of evolutionary algorithm that can be used for hardware evolution, there is an effect whereby the phenotype circuits produced tend to be relatively unaffected by small amounts of mutation to their genotypes. This effect can be turned to engineering use, such as encouraging parsimonious solutions or giving a degree of graceful degradation in the presence if certain hardware faults. There are other mechanisms by which evolution can be more explicitly induced to produce fault-tolerant circuits.

The DARPA (Defense Advanced Research Projects Agency) sponsored a Grand Challenges on Oct 8, 2005, which aimed to develop high-performance unmanned robotic vehicles which are capable of successfully driving though a long-distance hostile environment and reaching the designated destination. The team that most quickly finished a designated route across the Southwest's Mojave Desert can win a $2 million prize. The fault-tolerant capability of an autonomous vehicles is especially important for such a tough task, since it is possible that component faults or failures may occur or even become unavoidable in navigating such a desert course.

9.2.3 *Swarm Robotics*

Swarm robotics intends to coordinate the robotic system, which is made up of a large number of simple physical robots, in order to accomplish the task which is hard to achieve by a single robot. The approach is inspired by the collective behavior exhibited by the social insects. The collective behavior emerges from the inter-robot interactions as well as the interactions between robots and their physical environments. It is a type of swarm intelligence applications, where the complex swarm behavior may be achieved from many primitive individual behaviors through continuous individual interactions and communications. Swarm robotic systems can be applied to certain demanding applications such as large-scale distributed sensing systems, miniaturization measurement systems, and low-cost systems. The focal point in the swarm robotics research lies in the self-organization and emergence issues. Distinguished from traditional single robotic systems, cooperation between robots in a swarm of robots are extremely important. Each robot is equipped with sensing devices, communication mechanisms, decentralized control schemes, and so forth. Their close cooperation may result in the accomplishment of more challenging tasks. Also, we should notice that the system robustness and adaptability is significantly increased for a swarm of robots, with respect to a single robot. With the constant flow of new technologies such as ubiquitous computing, embedded devices, and wireless communications, swarm robotics is expected to become more realizable in real-world environments in the near future.

Overall, the continuous occurrence of novel technologies including improved hardware devices and more efficient control algorithms will definitely provides new opportunities in intelligent robotics development. It is believed that some of the fancy science-fiction stories will become true one day.

Index

Printed in the United States
By Bookmasters